Lecture Notes in Computer Science 11528

More information about this series at http://www.springer.com/series/7408

Andre Coy · Yugo Hayashi ·
Maiga Chang (Eds.)

Intelligent Tutoring Systems

15th International Conference, ITS 2019
Kingston, Jamaica, June 3–7, 2019
Proceedings

Springer

Editors
Andre Coy
University of the West Indies
Kingston, Jamaica

Yugo Hayashi
Ritsumeikan University
Osaka, Japan

Maiga Chang
Athabasca University
Edmonton, AB, Canada

ISSN 0302-9743 ISSN 1611-3349 (electronic)
Lecture Notes in Computer Science
ISBN 978-3-030-22243-7 ISBN 978-3-030-22244-4 (eBook)
https://doi.org/10.1007/978-3-030-22244-4

LNCS Sublibrary: SL2 – Programming and Software Engineering

This Springer imprint is published by the registered company Springer Nature Switzerland AG
The registered company address is: Gewerbestrasse 11, 6330 Cham, Switzerland

Preface

The 15th International Conference on Intelligent Tutoring Systems (ITS 2019) was held in Kingston, Jamaica, during June 3–7, 2019, under the auspices of the Institute of Intelligent Systems (IIS).

The theme of ITS 2019 was "Challenges in Artificial Intelligence Pushing the Boundaries of ITS" with the objective to present ongoing scientific contributions and show the impact of the ITS conferences on the field of intelligent systems in education and across other disciplines. The conference emphasized the use of advanced computer technologies and interdisciplinary research for enabling, supporting, and enhancing human learning. It promoted high-quality interdisciplinary research creating a forum for presenting and sharing challenges and novel advancements in artificial intelligence. It triggered an exchange of ideas in the field, reinforcing and expanding the international ITS network of researchers, academics, and market representatives.

The main track was composed of full and short scientific research works that represented the research results in using advanced computer technologies and interdisciplinary research for enabling, supporting, and enhancing human learning. The conference also included a poster track providing an interactive forum for authors to present research prototypes to the conference participants, as well as work in progress.

The international Program Committee consisted of 100 leading members (25 senior and 75 regular) of the intelligent tutoring systems community. The conference (general) chair was Andre Coy from the University of the West Indies, Jamaica; the Program Committee chairs were Maiga Chang from Athabasca University, Canada, and Yugo Hayashi from Ritsumeikan University, Japan.

Research papers were reviewed by at least three reviewers (with the majority receiving five or more reviews) through a double-blind process. Only 33.33% of papers submitted as full papers were accepted; 13 were accepted as short papers with six pages published in the proceedings. Four posters presentations that were directly submitted to the poster track chaired by Sabine Graf (Athabasca University, Canada) were also accepted. We believe that the selected full papers describe some very significant research and the short papers some very interesting new ideas, while the posters presented research in progress that deserves close attention.

Additionally, the ITS 2019 program included the following workshops and tutorials selected by the workshop and tutorial chairs, Ted Carmichael from University of North Carolina, USA, and Vasiu Radu from Polytechnic University of Timisoara, Romania:

- Optimizing Human Learning: Second Workshop Eliciting Adaptive Sequences for Learning (WeASeL 2019)
- AutoTutor Tutorial: Conversational Intelligent Systems and Learning Analytics
- Artificial Intelligence Code Tutorial
- Engaging modern learners: Shaping Education with Tablets in Schools
- Innovations at the University of West Indies related to AI, ITS, and Learning Systems

Furthermore, a panel discussion was organized on the topic of " Intelligent Tutoring Systems vis a vis Developments in Education and Industry." Finally we had an outstanding invited speaker, Amruth N. Kumar from Ramapo College of New Jersey, USA, who presented the topic: "Fifteen Years of Developing, Evaluating and Disseminating Programming Tutors: Lessons Learned."

In addition to the above contributors, we would like to thank all the authors, the various conference chairs, the members of the Program Committees of all tracks, the Steering Committee members, and in particular the chair, Claude Frasson. We would also like to acknowledge the conference organizer NEOANALYSIS and especially Kitty Panourgia and her excellent team for the permanent follow-up of the organization. We are also grateful to the conference hosting institution, the University of the West Indies, for all the scientific, organization and practical contributions.

Last but not least, we express our gratitude to the conference sponsors, in particular Springer for the Best Paper Award, as well as the Vice Chancellor of the University of the West Indies, and e-Learning Jamaica Company (e-LJam), the e-learning Agency of the Ministry of Science, Energy and Technology (MSET) in Jamaica, for their financial support.

June 2019

Andre Coy
Maiga Chang
Yugo Hayashi

Chairs/Committees

Conference Chair

Andre Coy	University of the West Indies, Jamaica

Program Chairs

Maiga Chang	Athabasca University, Canada
Yugo Hayashi	Ritsumeikan University, Japan

Workshop and Tutorial Chairs

Ted Carmichael	University of North Carolina, USA
Vasiu Radu	Polytechnic University of Timisoara, Romania

Poster and Interactive Events Chairs

Sabine Graf	Athabasca University, Canada
Gunjan Mansingh	University of West Indies, Jamaica

Award Chair

Stefano A. Cerri	LIRMM: University of Montpellier and CNRS, France

Doctoral Consortium Chair

Ricardo Anderson	University of the West Indies, Jamaica

Industry Track Chairs

Stephen Fancsali	Carnegie Learning, Inc., USA
Amruth Kumar	Ramapo College, USA

Volunteers Chair

Daniel Fokum	University of the West Indies, Jamaica

Organization Chair

Kitty Panourgia	Institute of Intelligent Systems (IIS), Greece

Senior Program Committee

Benedict Du Boulay	University of Sussex, UK
Bert Bredeweg	University of Amsterdam, The Netherlands
Beverly Park Woolf	University of Massachusetts, USA
Claude Frasson	University of Montreal, Canada
Demetrios Sampson	University of Piraeus, Greece
Esma Aimeur	University of Montreal, Canada
Gilles Gauthier	UQAM, Canada
James Lester	North Carolina State University, USA
Jean-Marc Labat	Université Paris 6, France
Julita Vassileva	University of Saskatchewan, Canada
Kevin Ashley	University of Pittsburgh, USA
Kinshuk Kinshuk	University of North Texas, USA
Maiga Chang	Athabasca University, Canada
Michaela Cocea	University of Portsmouth, UK
Michel Desmarais	Ecole Polytechnique de Montreal, Canada
Peter Dolog	Aalborg University, Denmark
Riichiro Mizoguchi	Japan Advanced Institute of Science and Technology, Japan
Roger Azevedo	University of Central Florida, USA
Roger Nkambou	University of Québec in Montréal (UQAM), Canada
Susanne Lajoie	McGill University, Canada
Stefan Trausan-Matu	Politehnica University of Bucharest, Romania
Stefano A. Cerri	LIRMM: University of Montpellier and CNRS, France
Tak-Wai Chan	National Central University, Taiwan
W. Lewis Johnson	Alelo Inc., USA
Yugo Hayashi	Ritsumeikan University, Japan

Program Committee

Ahmed Tlili	Beijing Normal University, China
Akihiro Kashihara	University of Electro-Communications, Japan
Alexandra Cristea	Durham University, UK
Alexandra Poulovassilis	University of London, UK
Amruth Kumar	Ramapo College of New Jersey, USA
Ashok Goel	Georgia Institute of Technology, USA
Benjamin Goldberg	United States Army Research Laboratory, USA
Blair Lehman	Educational Testing Service, USA
Carla Limongelli	University of Rome 3, Italy
Chih-Kai Chang	National University of Tainan, Taiwan
Chao-Lin Liu	National Chengchi University, Taiwan
Chih-Yueh Chou	Yuan Ze University, Taiwan
Cyrille Desmoulins	Université Joseph Fourrier, France
Darina Dicheva	Winston-Salem State University, USA
Davide Fossati	Carnegie Mellon University, Qatar

Diego Dermeval	Federal University of Alagoas, Brazil
Dunwei Wen	Athabasca University, Canada
Elise Lavoué	University of Lyon, France
Elvira Popescu	University of Craiova, Romania
Emmanuel Blanchard	IDÛ Interactive Inc., Canada
Éric Beaudry	UQAM, Canada
Evandro Costa	Federal University of Alagoas, Brazil
Fabio Akhras	Renato Archer Center of Information Technology, Brasil
François Bouchet	Sorbonne Université – LIP6, France
Fuhua Oscar Lin	Athabasca University, Canada
Galia Angelova	Bulgarian Academy of Science, Bulgaria
Genaro Rebolledo-Mendez	University of Veracruz, Mexico
Gwo-Jen Hwang	National Taiwan University of Science and Technology, Taiwan
Imène Jraidi	University of Montreal, Canada
Ivon Arroyo	Worcester Polytechnic Institute, USA
Jason Harley	University of Alberta, Canada
Kazuhisa Miwa	Nagoya University, Japan
Kuo-Chen Li	Chung-Yuan Christian University, Taiwan
Kuo-Liang Ou	National Hsin-Chu University of Education, Taiwan
Li Wang	Open University of China, China
Maher Chaouachi	McGill University, Canada
Maria Bielikova	Slovak University of Technology in Bratislava, Slovakia
Maria Lucia Barron-Estrada	Instituto Tecnologico de Culiacan, Mexico
Mark Core	University of Southern California, USA
Mark Floryan	University of Virginia, USA
Mary Jean Blink	TutorGen, Inc., USA
Michelle Taub	North Carolina State University, USA
Min Chi	North Carolina State University, USA
Mohammed Abdel Razek	King Abdulaziz University, Saudi Arabia
Nathalie Guin	University of Lyon 1, France
Nguyen-Thinh Le	Humboldt Universität zu Berlin, Germany
Nian-Shing Chen	National Sun Yat-sen University, Taiwan
Nicola Capuano	University of Salerno, Italy
Olga C. Santos	aDeNu Research Group, UNED, Spain
Patricia Jaques	UNISINOS, Brazil
Philippe Dessus	Université Grenoble Alpes, France
Reva Freedman	North Illinois University, USA
Riccardo Mazza	University of Lugano/University of Applied Sciences of Southern Switzerland, Switzerland
Rita Kuo	New Mexico Institute of Mining and Technology, USA
Robert Farrell	IBM, USA
Rod Roscoe	Arizona State University, USA
Sabine Graf	Athabasca University, Canada

Scotty Craig	Arizona State University, USA
Seiji Isotani	University of Sao Paulo, Brazil
Stephen B. Blessing	University of Tampa, USA
Tassos Mikopoulos	University of Ioannina, Greece
Tatsunori Matsui	Waseda University, Japan
Ted Carmichael	University of North Carolina at Charlotte, USA
Thepchai Supnithi	National Electronics and Computer Technology Center, Thailand
Valéry Psyché	TÉLUQ University, Canada
Vasiu Radu	Politechnica University of Timisoara, Romania
Vive Kumar	Athabasca University, Canada
Xiaoqing Gu	East China Normal University, China
Yusuke Hayashi	Hiroshima University, Japan
Carolina Mejia Corredor	Universidad EAN, Colombia
Marco Temperini	Sapienza University of Rome, Italy
Charalampos Karagiannidis	University of Thessaly, Greece
Silvia Margarita Baldiris Navarro	Fundación Universitaria Tecnológico Comfenalco, Colombia
Cecilia Avila	Universitat de Girona, Spain
Ting-Wen Chang	Beijing Normal University, China

Organizing Committee (NEOANALYSIS)

Isaak Tselepis (web architect)
Mara Gassel
Alexia Kakourou
Dimitris Sevastakis
Elissavet Vasileiou

15 Years of Developing, Evaluating, and Disseminating Programming Tutors: Lessons Learned (Invited Talk)

Amruth N. Kumar

Ramapo College of New Jersey, Mahwah NJ 07430, USA
amruth@ramapo.edu

Abstract. The past can inform the future when it comes to pushing the boundaries of ITS. Based on the experience of developing, evaluating, and disseminating two suites of software tutors for computer programming, viz., problets and epplets, I would like to proffer some lessons learned. Among those are: How correct is a pragmatic alternative to why incorrect?; Learning gains are not always commensurate with development costs; An ounce of direction is worth a pound of correction; Can do is not the same as should do; Solving ill-defined problems is about knowing what to ask and when; Mastery learning assessed in terms of effort rather than correctness; All that glitters in the laboratory may not be gold in the field; One size does not fit all; The path of least resistance can waylay the best of intentions; Learning is a whole person activity; When you are given lemons, make lemonade; Do it right *and* do it twice; If you build it, they will not come; and Dissemination is a Sisyphean task!

Speaker Bio: Amruth Kumar (PhD in Computer Science from the University at Buffalo) is Professor of Computer Science at Ramapo College of New Jersey, Mahwah, NJ, USA. His research interests include intelligent tutoring systems, educational data mining, computer science education research, and computer science education. He is the developer of problets (problets.org) and epplets (epplets.org) – two software tutoring suites on computer programming. His research has been funded by several grants from the National Science Foundation. He is a Distinguished Member of the ACM and Senior Member of IEEE.

Contents

A Learning Early-Warning Model Based on Knowledge Points

Jiahe Zhai, Zhengzhou Zhu(✉), Deqi Li, Nanxiong Huang, Kaiyue Zhang, and Yuqi Huang

School of Software and Microelectronics, Peking University, Beijing, People's Republic of China
zhuzz@pku.edu.cn

Abstract. Learning early-warning is one of the important ways to realize adaptive learning. Aiming at the problem of too large prediction granularity in learning early-warning, we divide student's characters into three dimensions (knowledge, behavior and emotion). Secondly, we predict the student's master degree of knowledge, based on the knowledge point. And then we realized learning early-warning model. In the model, we take 60 points as the learning early-warning standard, and take RF and GDBT as base classifiers, and give the strategy of selecting the basic model. The experiment shows that the prediction of knowledge mastery of the model and the real data Pearson correlation coefficient can reach 0.904279, and the prediction accuracy of the model below the early-warning line can reach 76%.

Keywords: Learning early-warning · Emotion · Type of question · Knowledge points

1 Introduction

Learning early-warning is to warn before the disadvantageous academic events, to avoid the harm caused by unexpected events, thereby reducing the loss caused by the harm. Learning early-warning is an optimum way to optimize teaching effect, realize teaching according to aptitude, and achieve the effect of individualized teaching. Learning early-warning can summarize a set of mechanisms, from data acquisition, performance prediction to early-warning information display [1]. How to make full use of learning data to build an effective model is one of the important problems in learning early-warning.

2 Related Works

The research and practice of learning early-warning at home and abroad are still in the initial stage, mainly the research of early-warning model. About the design of theoretical framework, Liu compared and summarized the existing learning early-warning system, proposed a learning early-warning framework, from data acquisition, performance prediction to early-warning information display, which has certain guiding

A. Coy et al. (Eds.): ITS 2019, LNCS 11528, pp. 1–6, 2019.
https://doi.org/10.1007/978-3-030-22244-4_1

significance [1]; in the empirical aspect, based on learning management system (LMS) in the analysis of learning records. Using unsupervised learning model prediction, Sabina based on Moodle learning platform log behavior records, through K-Means clustering to classify students into different categories, analysis of different types of students to fail [2]; also using supervised learning model prediction. For the labeled case, the early-warning content has the regression problem aiming at academic achievement. Cheng et al. designed a learning early-warning model based on Stochastic Forest according to the characteristics of knowledge, behavior and attitude based on online and offline behavior records [3]. There are also classifications for dropouts. Purdue University's curriculum information system has designed an early-warning model for dropouts [4].

This paper takes the second bachelor degree students of software engineering course as the research object, collects data through Moodle platform, filters, cleans and fuses data through feature engineering, and finally designs a learning early-warning model based on knowledge points and questions, and evaluates it through experiments.

3 Model Design

We extracts features from three dimensions of knowledge, behavior and emotion, which are mainly divided into basic information, learning basis, learning style, video viewing, daily testing and forum text emotion.

This paper designs a learning early-warning model based on knowledge points and problem types. We divide the predicted granularity into single knowledge points, and represent the mastery degree of a problem type according to the characteristics of different types of interrelated knowledge points, so as to obtain a representation of the final mastery degree. The basic framework of the model is stacked ensemble learning, i.e. multi-classifier system. In this framework, we can replace the adjustable base classifier by ourselves.

The model is divided into two phases showed in Fig. 1. In phase II, the knowledge points are divided into three types of questions to be trained separately. The feature vectors corresponding to the specific knowledge points are not treated differently in the same type of questions, and then the optimal feature combination training model is selected by features for the next stage. In the second phase, the classifier of the first stage is used to predict the knowledge points involved in different types of questions in the test paper. The average is obtained as the feature vector of the second stage training, and then the knowledge mastery degree of the test paper is predicted by direct training.

3.1 Phase I

(a) Constructing data sets: The data are divided into three data sets according to the type of questions. For different types of questions, the characteristic data of the same knowledge point are the same, but the difference is the score of knowledge points. Then, the data set is balanced according to the score of knowledge points, and divided into training set and test set in the proportion of 7:3. Different types of questions have unique training and test sets for training and prediction.

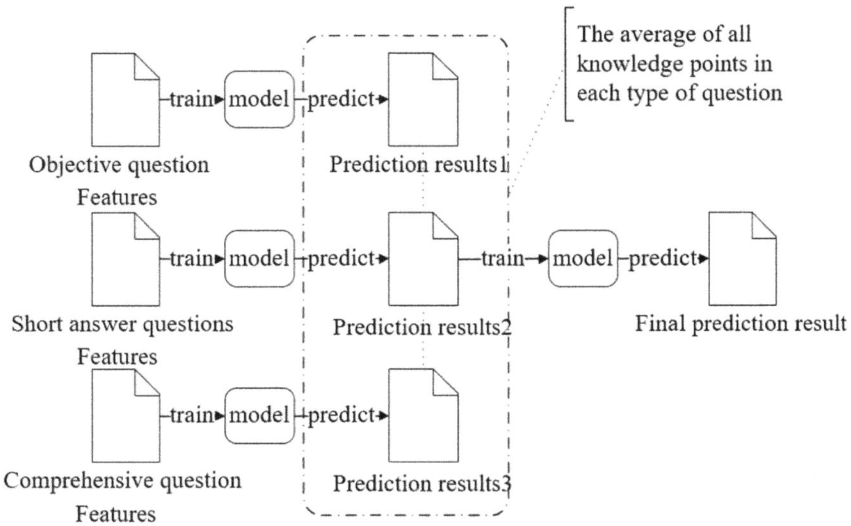

Fig. 1. Model framework

(b) Training model: Firstly, the importance of features is sorted by using base classifier in training set, and then the model is trained iteratively by searching feature combination. The model trained on training set is tested on test set, and the optimal feature combination is selected, and then the optimal model is selected.
(c) Prediction: Predict the test set with the model of the previous stage, and the result is the degree of mastery of the knowledge points of the corresponding questions.

3.2 Phase II

(a) Constructing data sets: The data sets in this stage are the predicted values of different types of questions in the previous stage, and the data sets are balanced with the final examination scores as the criteria. Considering that the data samples are relatively small, in order to make the best use of the data, this step uses the retention method to divide the data sets, that is, one sample data as the test set, and the rest as the training set.
(b) Training model: In this stage, the model is directly trained by using base classifier without feature selection.
(c) Prediction: Use the training model of the previous stage to forecast the samples one by one to form the prediction results.

(d) Determine the early-warning area: the model uses 60 as the early-warning standard, less than 60 as the required early-warning range, and more than 60 as the safe range.

Because of the characteristics of the question type, the model divides the question type into three parts: objective question, short answer question and comprehensive question. A model is trained for each of the three types of mid-term questions, and then the model is used to train the knowledge points of the final exam, and then the average is obtained. As the eigenvector of the next round of training, the final exam results are used as the final prediction label.

In this paper, we use wrapped feature selection, that is to say, we use the importance of feature after model training to sort, and the importance of feature ranking based on specific classifiers is the best trainer for classifiers. Selection strategy does not use backward search, we do not stop because the best time, but directly from the first to the last, although this cannot exhaust all feature combinations, but the effect must be better than the traditional forward search.

4 Data Analysis

As shown in Fig. 2, the distribution of the original fraction is not uniform, which is also normal and approximate to normal distribution. But when classifier makes regression calculation, it is through fitting residual. If a certain distribution proportion is too large, the fitting process of the model will deviate from this part of the data, because even if it deviates from a small part of the data, the total can still be obtained. For the small sum of squared variance, this paper makes a balanced sampling of the fraction interval of 5, and finally obtains that there are 221 learner samples in the test set.

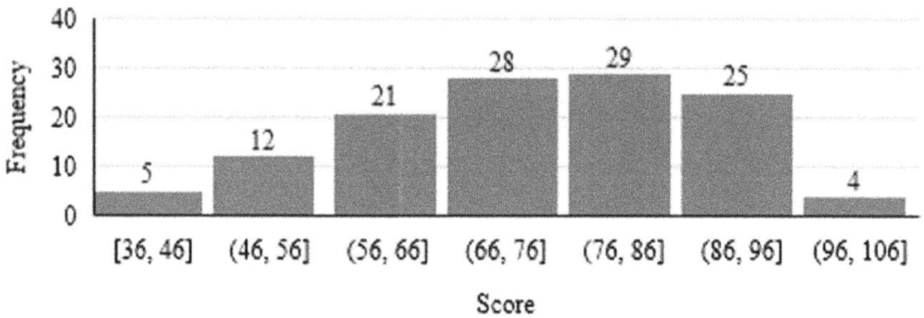

Fig. 2. Score distribution

The basic classifier of model training compares two representative classifiers, RF and GDBT. The results of performance prediction are compared as shown in Fig. 3. Considering the small number of samples in this paper, and in order to better characterize the performance of the model, we use the retention method.

As shown in Fig. 3, the trend of performance prediction is consistent. Pearson correlation coefficient is used in this paper.

As shown in Table 1, the RF's correlation is 0.880568 and 0.904279, respectively. The correlation is very large, so the prediction effect of this model is good.

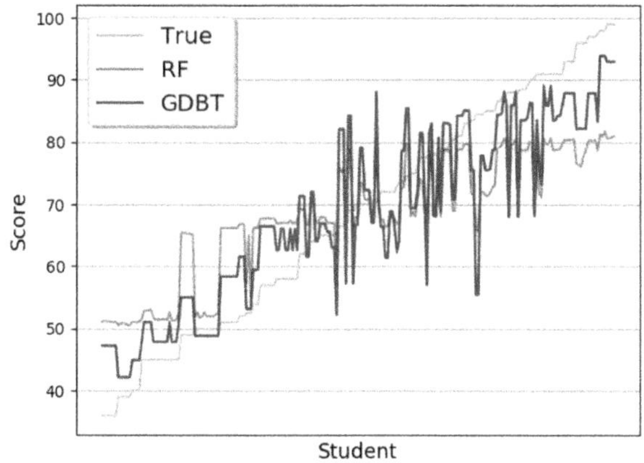

Fig. 3. Performance prediction comparison of 3 classifiers

Table 1. Predictive indicators for different classifiers

	Pearson corr	MAE	MSE	MSLE
RF	0.880568	9.281110	114.777580	0.026352
GDBT	0.904279	6.499410	65.149360	0.014421

Mean Absolute Deviation (MAE), Mean-Square Error (MSE) and Mean-Square Logarithm Error (MSLE) are also used to characterize the prediction of continuous values. These three characteristics are also used to compare other prediction methods. Of course, as can be seen from Fig. 4, the model fitting in this paper is still inadequate, which requires more detailed feature screening and larger samples to improve.

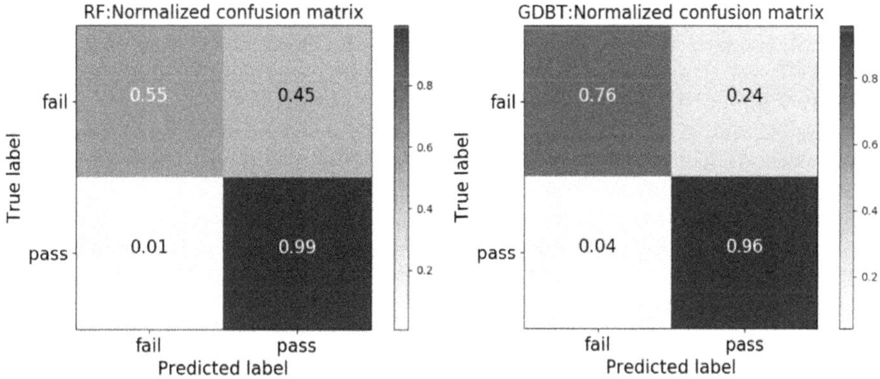

Fig. 4. Comparison of obfuscation matrices

The purpose of the model is to carry out learning early-warning, and it is not enough to predict the achievement alone. So 60 is divided into learning early-warning criteria, that is, the most general passing line. The corresponding confusion matrix is shown in Fig. 4. GDBT is better for the degree of sub-standard, with an accuracy rate of 0.76. For the above-standard prediction, the accuracy of RF is 0.03 higher than that of GDBT. First, the difference is slight. Second, considering the purpose of learning early-warning, we can tolerate setting students above the standard as early-warning targets, and at most urge them. At the same time, according to the confusion matrix, we can calculate F_β values of 0.638453 and 0.805529, respectively. GDBT is also significantly better than RF. In summary, this paper tends to choose GDBT as the base classifier.

5 Summary and Prospect

In this paper, a specific learning and early-warning model is trained based on the fine-grained partition of knowledge points and questions. We can put the data of a single knowledge point into the model and use three models to predict the level of a single knowledge point or the overall mastery, so as to achieve the early-warning effect. We will apply the early-warning algorithm to the next students in the next semester to help learners improve their performance.

Without discussion on feature intervention, users can adjust themselves according to the corresponding application scenarios, such as dividing the warning level and adopting corresponding intervention measures. The amount and type of data in this paper are limited. The next step is to expand the quantity and type of data.

Acknowledgments. This paper was supported by National Key Research and Development Program of China (Grant No. 2017YFB1402400), Ministry of Education "Tiancheng Huizhi" Innovation Promotes Education Fund (Grant No. 2018B01004), National Natural Science Foundation of China (Grant No. 61402020), and CERNET Innovation Project (Grant No. NGII20170501).

References

1. Liu, J., Yang, Z., Wang, X., Zhang, X., Feng, J.: An early-warning method on e-learning. In: Liu, S., Glowatz, M., Zappatore, M., Gao, H., Jia, B., Bucciero, A. (eds.) eLEOT 2018. LNICST, vol. 243, pp. 62–72. Springer, Cham (2018). https://doi.org/10.1007/978-3-319-93719-9_9
2. Cheng, X., et al.: A novel learning early-warning model based on random forest algorithm. In: Nkambou, R., Azevedo, R., Vassileva, J. (eds.) ITS 2018. LNCS, vol. 10858, pp. 306–312. Springer, Cham (2018). https://doi.org/10.1007/978-3-319-91464-0_32
3. Sisovic, S., Matetic, M., Bakaric, M.B.: Clustering of imbalanced moodle data for early alert of student failure. In: 2016 IEEE 14th International Symposium on Applied Machine Intelligence and Informatics (SAMI), pp. 165–170. IEEE Press, Herlany (2016). https://doi.org/10.1109/sami.2016.7423001
4. Purdue Information Technology. http://www.itap.purdue.edu/learning/tools/signals/

Adaptive Learning Spaces with Context-Awareness

Valéry Psyché[1], Ben Daniel[2(✉)], and Jacqueline Bourdeau[1]

[1] TÉLUQ University, 5800 rue Saint-Denis, Montréal, QC H2S 3L5, Canada
{valery.psyche, jacqueline.bourdeau}@teluq.ca
[2] Otago University, Dunedin, New Zealand
ben.daniel@otago.ac.nz

Abstract. This paper introduces the concept of learning space in the digital age and considers the various contexts in which learning occurs. In addition, the paper discusses issues about the modelling of a context-aware learning space ontology. The paper is based on an ongoing collaborative research, which focuses on the interface of learning space and context-aware learning. The long-term goal of this project is to introduce the notion of learning space and explore the role of the design of learning space context-aware ontologies with the ultimate aim of constructing a transformative theory of context-aware learning spaces such as personal learning networks, virtual learning spaces, social learning spaces, and cognitive learning spaces.

Keywords: Learning space · Learning context · Ontology · Context awareness · ITS

1 Introduction

Learning spaces are often depicted as classrooms with students seating in rows, listening and taking notes, with a teacher or lecturer standing in front of them, delivering knowledge or information. This model of learning space assumes that the student's progress toward a programme of study is determined by the time spent in classrooms, his place in the classroom and his interactions with teachers and other students. As such, the physical design and organisation of the classroom and the seating position of the student in the classroom can affect performance [1]. However, the changing landscape of learning environments and students (i.e. diversity in students and learning needs, and the permeation of digital technologies into learning), suggest that the conventional understanding of learning space, be it the formal lecture room, the seminar room or tutorial room, is untenable for all types of learning modalities of the 21st century.

The 21st-century students are social, team-oriented, multi-taskers who have a positive outlook on life. They are hands-on with a "let us build it" approach that places increasing value on network devices [2]. Further, learning for this generation has now become a lifelong pursuit, which takes within technological frontiers, supporting physical, online and blended [3, 4].

A. Coy et al. (Eds.): ITS 2019, LNCS 11528, pp. 7–13, 2019.
https://doi.org/10.1007/978-3-030-22244-4_2

The modern teacher is likely to be using various pedagogical approaches such as case-based, problem-based learning, community-oriented pedagogues; where the teacher assumes the role of a facilitator, and students work in groups and teams. [5] stated that the impact of digital technologies since the mid 1990s has implications for where and how learning might happen, whether it is online or offline, situated or distributed. It is also worth noting that while classrooms are formal learning spaces, distributed, and networked learning environments can take forms of informal and non-formal learning spaces.

In this paper, we first introduce the concept of learning space in the digital age, and the various contexts in which learning occurs. We then provide various dimensions of the concept of learning space, taking into account the context in which each dimension can support learning. We present our ideas for modelling learning spaces based on the work of [6]. Further, we discuss the issues of context and learning space in the light of a context-aware learning space framework. This framework will support various experiments on instructional design in order to improve adaptive and personal learning within the networked educational paradigm.

2 Related Research

Discussions of learning space within many higher education institutions largely remain constrained to three areas, namely the classroom (where almost all learning occurs), the library and the faculty offices—where programs are designed and student work graded [7]. Today, learning environments, as we know it, transcend physical spaces to virtual, cognitive and social spaces.

A growing body of research has called for the rethinking of learning spaces needed for the 21st century [7–9]. It is generally noted that research into learning space provides an opportunity to inform the development of adaptive and personalised technologies to enrich the student individual is learning need. Beyond that perspective, it is possible to open the learning space in order to support the establishment of a social network for exchanges between students across the planet, leading to new educational experiences that might otherwise not be possible to achieve. As [10] noted that social networking technology supplement face-to-face courses and can enhance students' sense of community.

Optimisation learning space requires the development of new learning paradigms that can adequately meet the needs of current generations of learners, the Net Generation (1982–1990s) and Gen Z (born 1995–2010) [2]. We take a holistic view of learning space, conceptualising it along fundamental dimensions (physical, virtual, social, and cognitive, see Fig. 1).

Virtual learning spaces comprise learning mediated both synchronous and asynchronous. In these environments, students learn to multitask and continually work outside of the classroom in spaces that promote social learning. Learning in the context of social networks is highly self-motivated, autonomous and informal and forms an integral part of the higher education experience [11]. Besides, social networks are considered useful in developing essential skills like selecting relevant information, critically interpreting and analysing the sociocultural context, working collaboratively

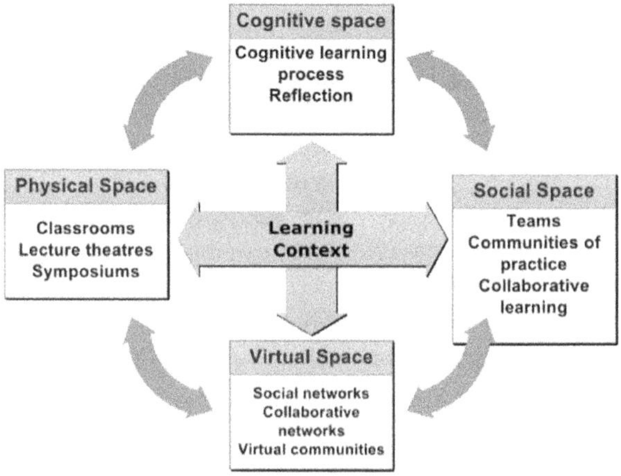

Fig. 1. Defining learning space.

and sharing knowledge [12]. In the social aspects of learning spaces, the concepts of learning ecology [4] and learning communities [13] are critical because they emphasise learning in a social context, recognising that learners are simultaneously involved in multiple learning settings.

A learning ecology is a collection of contexts—physical and virtual—that provides opportunities for learning [4]. In higher education, this usually includes multiple courses, formal and informal contexts across the institution, and settings at work, community and home. Social learning spaces are instrumental in setting conditions for learning because they create a supportive environment to engage students in critical thinking and promotes interactions that are richer, more gratifying and intrinsically rewarding [12].

Physical learning spaces are also valuable learning environments, and they are considered part of the aesthetic view, one of identity and symbolic of power and prestige. Beyond the classroom, physical learning spaces are quiet spaces or individual pods for individual or small groups; break out spaces that could be large or small and widened corridors allowing the gathering of students away from the formal learning environments.

3 Modelling Learning Space Inside Learning Context

In artificial intelligence, the notion of context appeared in the 1990s [14], but it was not until the early 2000s where this area of research gained interests among researchers in ubiquitous computing, mainly focusing on geolocalization technologies, where the spatial and temporal dimension of the context became traceable. [15] have analysed one hundred and fifty definitions of the context. From this study results the most cited model of context among research. It represents the components of a situation and the relations between them. According to Bazire and Brezillon “A situation could be defined by a user, an item in a particular environment, and eventually an observer." In their model in context, context and environment are separate but related.

According to [16], "space is an important context in many context-aware applications, and most context definitions mention space as a vital factor". [17] described "context as any information that can be used to characterise the situation of an entity, where an entity is a person, place or object that is considered relevant to the interaction between a user and an application". The notion of learning context in education describes the various circumstances in which learning might occur, where the learning context consists of students, culture, teachers, lesson plans, etc. In addition, learning context refers to the set of learning activities, including the space in which learning itself occurs, and students' perceptions of the ability of a particular space in providing rich learning experiences.

Knowledge of learning context enables both teachers and students to rethink about the design of teaching and learning, and the constraints of the learning spaces [18]. The affordances of a context must be perceived by an individual who must also have the abilities to interact with these attributes. Openness disrupts teaching conventions; however, it is the social activity of the inhabitants that define the possibilities of a learning space [19]. Moreover, learning context consists of students, culture teachers, lesson plans, etc.

The emergence of Massive Online Open Courses (MOOC) in 2008 and the subsequent possibility of accessing large data about student interactions in online learning situations triggered more interests in understanding context and learning. Recently, studies have emerged in teaching in context [6, 20]. [21] stressed the importance of the external context in networked collaborative learning. According to these authors, the external context of a learning situation is influenced by environmental factors that have subsequent impacts on the learning process.

4 Development of the Approach

In order to investigate the linkage between learning space and context, we will adopt the Design-Based Research (DBR) methodology, which considers the process of design as important as the product, and where, each iteration is considered a sub-result leading to the next one [22]. The methodology would involve the conception and modelling of the context model of different learning spaces. The model draws from [15], which considers various forms of the environment (physical, virtual, social and cognitive). In this model, we take into account the "spatiotemporal location" component of the context, where the "items" represents any learning systems (e.g. in intelligent tutoring, computer-supported collaborative learning systems or massive open online courses).

The second phase of the methodology will involve the construction of an ontology of learning spaces. This ontology will be built from a reflexion on the relationship between learning space and learning situation (see Table 1). The ontology will inform the development of use case scenarios demonstrating various forms of learning spaces.

Table 1. Examples of learning situation

Type of space	Learners	Learning	Teacher
Classroom	Group[a]	Individual	Human Tutor
Classroom	Team inside a group	Collaborative	Human Tutor and Facilitator
CSCL[b] System	Team or Group	Collaborative	Facilitator
ITS[c]	Learner	Individual	Intelligent Tutor
MOOC[d]	Learner inside a Massive Virtual Group	Individual	No Tutor/Facilitator

[a]A Group = around 30 learners; a Team = 3 to 6 learners, and Individual Learning = 1 to 3 learners.
[b]Computer Support for Collaborative Learning.
[c]Intelligent Tutoring System.
[d]Massive Online Open Course.

The third phase of the methodology would involve running user experiments, where learners will engage in learning activities, and their overall learning experiences will be evaluated.

The experimental design will take into account the possible factors they attribute to enhance learning outcomes or experiences, and space and context in which this occurs. Besides, the design of the learning activities will be informed by an instructional design model, which involves analysis, design, development, implementation and evaluation (ADDIE) [23]. We will also collect learning analytics and learners' profiles to build a knowledge base (activity trace templates) [24]. The knowledge base analysis will help validate the ontology of the learning spaces and the discovery of contextual knowledge.

5 Summary and Future Work

Digital learning technologies have transformed the way students engage and interact off and online, yet the physical learning spaces in which learning occurs has not changed much. In this paper, we introduce the concept of learning space and the various contexts in which learning might occur. This is work in progress; future work involves the development of context-aware ontologies and conducting a series of experiments to construct a transformative learning theory that takes into account the various contexts of learning spaces (physical, virtual, social, and cognitive).

We are aware that it is unlikely that providing support to all forms of learning spaces can necessarily enable students to transition from one space to another without facing any challenges. Therefore, there is a need to address out of class physical and virtual learning spaces to encourage learning is critical as [25] noted that students tend to spend more time in informal learning spaces rather than formal learning environments.

References

1. Yuan, Z., Yunqi, B., Feng-Kuang, C.: An investigation of university students' classroom seating choices. J. Learn. Spaces **6**, 13–22 (2017)
2. Brown, M.B., Lippincott, J.K.: Learning spaces: more than meets the eye. EDUCAUSE Q. **36**, 14–17 (2003)
3. Garrison, D.R.: E-learning in the 21st Century: A Framework for Research and Practice. Routledge, Abingdon (2011)
4. Scott, K.S., Sorokti, K.H., Merrell, J.D.: Learning "beyond the classroom" within an enterprise social network system. Internet High. Educ. **29**, 75–90 (2016)
5. Erstad, O.: The expanded classroom-spatial relations in classroom practices using ICT. Nord. J. Digit. Lit. **1**, 8–22 (2014)
6. Forissier, T., Bourdeau, J., Mazabraud, Y., Nkambou, R.: Modeling context effects in science learning: the CLASH model. In: Brézillon, P., Blackburn, P., Dapoigny, R. (eds.) CONTEXT 2013. LNCS (LNAI), vol. 8175, pp. 330–335. Springer, Heidelberg (2013). https://doi.org/10.1007/978-3-642-40972-1_25
7. Temple, P.: Learning spaces in higher education: an under-researched topic. Lond. Rev. Educ. **6**, 229–241 (2008)
8. Osborne, J.: The 21st century challenge for science education: assessing scientific reasoning. Think. Ski. Creat. **10**, 265–279 (2013)
9. Selwyn, N.: Social media in higher education. Eur. World Learn. **1**, 1–10 (2012)
10. Hung, H.T., Yuen, S.C.Y.: Educational use of social networking technology in higher education. Teach. High. Educ. **15**, 703–714 (2010)
11. Dabbagh, N., Kitsantas, A.: Personal Learning Environments, social media, and self-regulated learning: a natural formula for connecting formal and informal learning. Internet High. Educ. **15**, 3–8 (2012)
12. Garrison, D.R., Anderson, T., Archer, W.: Critical thinking, cognitive presence, and computer conferencing in distance education. Am. J. Distance Educ. **15**, 7–23 (2001)
13. Schwier, R.A., Morrison, D., Daniel, B.K.: A preliminary investigation of self-directed learning activities in a non-formal blended learning environment (2009)
14. McCarthy, J.: Notes on formalizing context. In: 13th International Joint Conference on Artificial Intelligence, pp. 555–560. Morgan Kauffman, San Francisco (1993)
15. Bazire, M., Brézillon, P.: Understanding context before using it. In: Dey, A., Kokinov, B., Leake, D., Turner, R. (eds.) CONTEXT 2005. LNCS (LNAI), vol. 3554, pp. 29–40. Springer, Heidelberg (2005). https://doi.org/10.1007/11508373_3
16. Bettini, C., et al.: A survey of context modelling and reasoning techniques. Pervasive Mob. Comput. **6**, 161–180 (2010)
17. Dey, A.K., Abowd, G.D., Salber, D.: A conceptual framework and a toolkit for supporting the rapid prototyping of context-aware applications. Hum.-Comput. Interact. **16**, 97–166 (2001)
18. Alterator, S., Deed, C.: Teacher adaptation to open learning spaces. Issues Educ. Res. **23**, 315 (2013)
19. Lefebvre, H.: The Production of Space, Trans. Donald Nicholson-Smith. Blackwell, Oxford (1991)
20. Psyché, V., Anjou, C., Fenani, W., Bourdeau, J., Forissier, T., Nkambou, R.: Ontology-based context modelling for designing a context-aware calculator. In: Workshop on Context and Culture in Intelligent Tutoring System at ITS 2018 (2018)

21. Bourdeau, J., Forissier, T., Mazabraud, Y., Nkambou, R.: Web-based context-aware science learning. In: 24th International Conference on World Wide Web Companion, pp. 1415–1418. Association for Computer Machinery (ACM) (2015)
22. Bourdeau, J.: The DBR methodology for the study of context in learning. In: Brézillon, P., Turner, R., Penco, C. (eds.) CONTEXT 2017. LNCS (LNAI), vol. 10257, pp. 541–553. Springer, Cham (2017). https://doi.org/10.1007/978-3-319-57837-8_44
23. Branch, R.M.: Instructional Design: The ADDIE Approach. Springer, Heidelberg (2009). https://doi.org/10.1007/978-0-387-09506-6
24. Daniel, B.K. (ed.): Big Data and Learning Analytics in Higher Education. Springer, Switzerland (2016). https://doi.org/10.1007/978-3-319-06520-5
25. Johnson, N.F.: The Multiplicities of Internet Addiction: The Misrecognition of Leisure and Learning. Routledge, Abingdon (2016)

An Adaptive Approach to Provide Feedback for Students in Programming Problem Solving

Priscylla Silva[1,3](✉), Evandro Costa[2], and Joseana Régis de Araújo[3]

[1] Federal Institute of Alagoas, Rio Largo, Brazil
priscylla.sousa@ifal.edu.br
[2] Computing Institute, Federal University of Alagoas, Maceió, Brazil
evandro@ic.ufal.br
[3] Federal University of Campina Grande, Campina Grande, Brazil
joseana@dsc.ufcg.edu.br

Abstract. This paper describes an approach to help students involved in a Programming Tutoring System, providing them with feedback during the coding problem-solving activities. It provides feedback for students during the coding, helping them to fix mistakes and how to take the next steps to complete the solution. This way, the student does not need to complete and submit a solution to get feedback from the system. The approach uses three feedback resources: videos, text hints, and flowcharts. We conducted an experiment involving 34 students from a programming introduction course. Preliminary results indicated a positive impact on the students learning. Our results also suggested that we can provide valuable feedback to students with difficult to complete a solution.

Keywords: Feedback · Programming education · Intelligent tutoring systems

1 Introduction

The skill of computer programming is one of the fundamental skills in courses in the field of computing. Moreover, particularly, many countries also include programming teaching in primary and high schools [11].

Many tools have been developed since 1960 with the goal of assisting the student by providing an environment for solving programming problems [4].

According to [9], there are two reasons to develop tools to help students in their programming learning process. The first reason is that programming learning is challenging, the students need help to progress in their learning, and they need constant practice. The second reason is that programming courses have many students and they need individually support, and the practice of programming requires much time of the teachers to provide feedback for the students [9].

To provide individualized support, many tools provide automatic feedback. In general, such tools are systems in which the students code solutions to problems and receive feedback message from the system informing whether their solutions are correct or not. This is the most common type of feedback found in these systems, which indicates only if the solution is correct based on a set of test cases [7]. This type of feedback is provided only after the student completes a solution and submits it for

A. Coy et al. (Eds.): ITS 2019, LNCS 11528, pp. 14–23, 2019.
https://doi.org/10.1007/978-3-030-22244-4_3

evaluation. Currently, there are few systems that provide student assistance during the coding of the solution in the programming context [6].

Proposals for these systems still present difficulties in providing adequate and useful feedback to students. According to [7], most systems provide low-quality feedback in a survey conducted with programming students, it was realized that most students can not submit a correct solution in their first attempt, and of these, few try a second time. In [7], they suggest that richer and detailed feedback be offered to the student, in this case, a possible solution is to provide feedback during the process of developing the solution, such as tips and step-by-step suggestions.

Providing feedback messages to the student to assist in coding a solution is not a trivial task. If the feedback is not adequate, the student may ignore any help proposed by the system. The feedback message should also be consistent with the current state of the student's solution. Also, many educational resources can compose the content of feedback such as video lessons, textual hints, model solutions, an explanation of the problem, simulations, and others [8].

The growing number of programming learning systems has produced new research opportunities as well as new challenges. One of the challenges is how to provide adaptive and automatic feedback.

According to [6], most existing feedback solutions for the programming domain offer post-submission feedback, i.e. awaiting the student complete the solution of a problem to provide a feedback message. According to [7], this type of feedback is not helpful for most students who have difficulties in the coding of the solution.

This paper presents an approach to provide adaptive feedback during the resolution of programming exercises. Feedback messages contain helping them to fix mistakes and how to take the next steps to complete the solution. Feedback messages have three formats: text, video, and flowcharts.

This paper is organized as follows, in the next section, some related works are shown. In Sect. 3, we present how the tutoring programming system works with immediate feedback during problem-solving. Our approach to providing adaptive feedback is described in Sect. 4. In Sect. 5 we present the execution of an experiment to verify the impact of our approach on student learning. Finally, the last section presents the conclusions and further work.

2 Related Work

In the [3], the authors make a systematic review of intelligent tutoring systems for programming education. They present 14 systems of which 12 contains solutions for adaptive feedback. Most systems created to assist students in programming learning offer only immediate feedback after a valid solution is submitted. Such systems do not have mechanisms that assist the student in the during the elaboration of the solution [6].

Provide support during the coding process is essential because situations can occur in which the student can not finalize a solution or can not even initiate a solution [7]. These situations can occur due to several factors, such as difficulty in understanding the problem statement, lack of knowledge of the programming topics required for the solution, difficulty in formalizing thinking in code, and others.

Few papers in the literature have as a proposal to provide feedback to the student during the coding process. In the [6], the authors analyzed 101 systems for programming teaching with automatic feedback generation, of this 32.7% generate feedback messages for error correction and only 18.8% generate messages that help the student how to proceed in the next steps.

The work of Corbett and Anderson [2] was one of the first to analyze the use of feedback in programming teaching systems. They used three types of feedback content: error-indicating messages, tips, and explanations. His approach used model solutions and was integrated into a system for teaching Lisp (a functional programming language).

The work presented in [5] describes a prototype of a tutor system for interactive programming. Each problem registered in the system is associated with one or more model solutions, created specifically with different problem-solving strategies. A human instructor notes model solutions and specific feedback messages can be associated with parts of the model solution code.

The work of Rivers and Koedinger [10] presents ITAP (Intelligent Teaching Assistant for Programming). This system uses historical data from exercise solutions to automatically generate customized tips for students. The system analyzes the current state of the student solution and the solution building path.

3 System with Adaptive Feedback

We developed an intelligent tutoring system for the students to practice programming with coding questions. The system provides a feedback approach to help students during the coding, helping them to fixed mistakes and how to take the next steps to complete the solution.

In our system, the students can use the C/C++ language for programming. The adaptive feedback is based on the status of the student' solution. When the student is solving a question, the system analyzes the code produced by the student. Whether the student requests help, the system selects the feedback message based on the analysis of the code, even if the code is not complete or syntactically correct. Currently, our system presents three types of feedback: text hints, videos, and flowcharts.

The system has a set of coding questions in which the student should write the code that solves the problem presented in each question. The system provides an interface for the student to write the code, ask for help, and submit the solution.

When students ask for help, they can choose the type of feedback they want. There is a button on the interface for each type of feedback, and a button for the student to visualize all the feedback messages received for the current question. Figure 1 shows the question resolution interface of the system.

The description of the question is visible at the top of the interface (noted A in Fig. 1). Bellow of that, there is the field for the student write the solution (B in Fig. 1). The options to request feedback are visible in the right. The student may indicate a specific part of the solution for which he wishes to help. He needs to indicate the lines in which the part of the code begins and ends (C in Fig. 1). This part is optional, and if the student does not indicate the lines, the system will use all the code of the solution to select a feedback message. The student selects the button of the type of feedback, and the selected message is displayed below of it (D in Fig. 1).

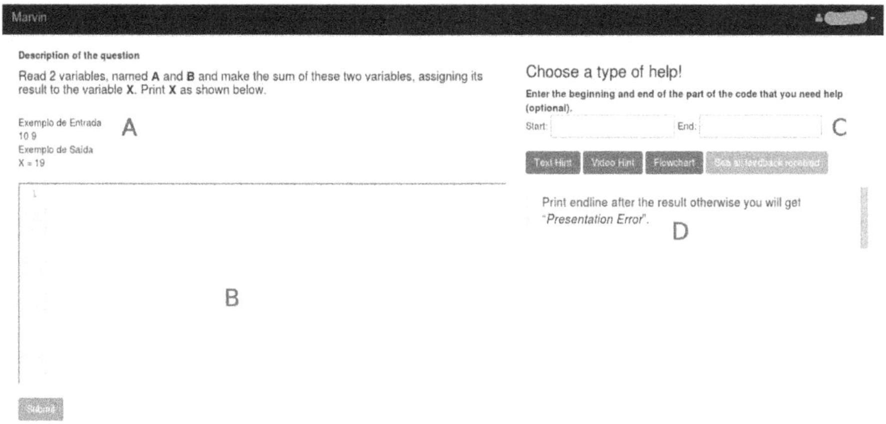

Fig. 1. Annotated screenshot of the system interface with adaptive feedback.

4 Approach to Provide Adaptive Feedback

In this section, we discuss our approach to select the feedback message most appropriate for the student.

To perform the feedback selection, we used model solutions registered by the teacher in the system. During the registration of the model solution, the teacher can associate feedback messages to different parts of the solution or the whole solution. The teacher can create feedback messages in three formats: text, video, and flowchart. One question may have several model solutions associated with it.

At the moment the student requests help, the system begins the process of selecting and presenting an adequate feedback message. Initially, the system checks if the student has told which part of the solution he/she needs help with it. If no part of the code has been specified, the system will use the current state of the student solution to search for general feedback. General Feedbacks are feedback messages whose content has been created and associated with a complete model solution.

If the student has specified the part of the solution for which he needs help, then the system will look for specific feedback. Specific feedbacks are messages whose content has been created and associated with code parts of model solutions.

The system performs the steps described in Fig. 2 to retrieve the most appropriate feedback message for the current state of the student solution.

Initially, the system performs the following steps to retrieve a list of appropriate feedback messages (search process):

– The system retrieves all feedback messages linked to the question that the student is currently solving;

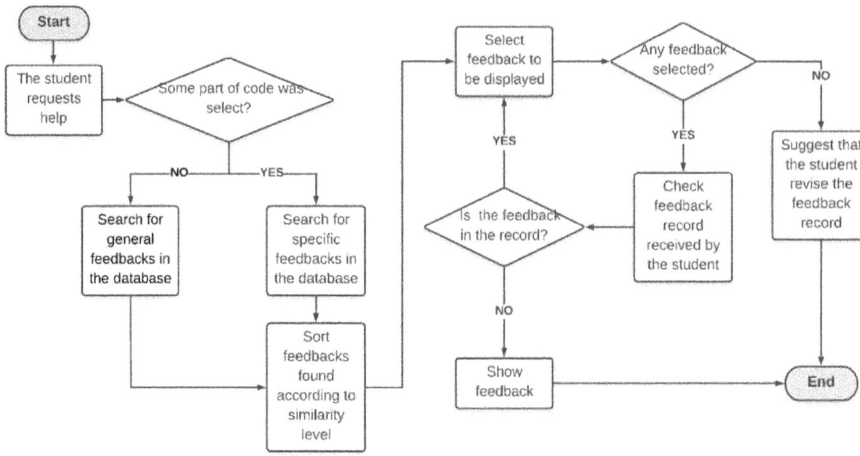

Fig. 2. Workflow process of selection of the feedback.

- The system extracts the part of the code specified by the student from the current state of his/her solution. If no part is specified, the entire solution will be extracted;
- The system extracts the code parts of the model solutions whose feedback messages are associated. In the case of general feedback, the entire solution is extracted;
- The system performs a similarity calculation between the part of the student solution code and the parts extracted from the model solutions; and
- All feedback messages whose similarity calculation is different from zero are returned.

After the search process, the system creates a list of the feedback messages returned in descending order according to the level of similarity. In the sequence, the first message is selected and compared to the record of messages already received by the student for the current question. If the message has already been presented to the student during the resolution of the question, the next message in the list is selected, and the process is repeated until a new message is displayed to the student, or if the student has already received all the feedback messages contained in the list, the system will recommend that the student review his or her message record.

The system uses a similarity calculation to determine which feedback message should be presented to the student. This similarity calculation is performed between the parts of the code provided by the student and the code parts of the model solutions. For the realization of this calculation, an abstract syntactic tree (AST) is generated for the model solution codes and the student solution code. An AST is a representation of the structure of the source code written in a programming language. The AST is serialized in string format. Then, the strings are compared using Levenshtein's distance algorithm. The comparison that results in the least number of modifications is considered the most similar.

5 Methodology of Evaluation

In this section, we discuss the evaluation of the approach to providing feedback during the coding problem-solving. We conducted a quasi-experiment to determine whether the approach has a positive impact on the student learning experience.

5.1 Participants

This study involves 34 participants (13 female, and 21 males). They are students from an introductory programming course in Brazil. The students were from the same class, and this was the first programming course of all them. They were aged between 17 and 30 years.

5.2 Experiment Design

We conducted a quasi-experiment to evaluate our approach on the 34 participants. The learning impact was evaluated by a quasi-experiment using a pre-test/post-test approach: all thirty-four students answered programming exercises, before and after the experience of using our feedback approach. This procedure was divided into three sessions. During the first session (60 min), the students answered the pre-test. Next, during the second session, they used the programming tutoring system with our feedback approach individually on a desktop computer (120 min). The students were instructed to answer three previously selected programming questions on the system, and if they needed help with the questions, they could only use the feedback provided by the system. In the end, in the third session (60 min), they were given the post-test.

The pre-test and post-test were composed of one programming question in which the students should applied basic programming constructs: sequence, conditions, and loops. The questions in the tests were evaluated using test cases and the scores determinate between zero to ten.

5.3 Results and Discussion

The students' performance in the post-test (M = 5.80, Mdn = 6.25, SD = 3.06) was better than in the pre-test (M = 4.58, Mdn = 5.00, SD = 2.77). The students' scores are presented in Fig. 3. These results are providing good evidence that the use of our feedback approach aided the students to improve them programming learning. To evaluate whether these results are significant, we conduct statistical tests. We used an alpha level of .05 for all tests.

First, we used the Shapiro-Wilk test to verify if the samples had a normal distribution. Both pre-test ($p = .08$) and post-test ($p = .079$) have a normal distribution. Then, a Paired T-Test was used to assess whether the difference between pre and post-test was significant.

The t-test applied resulted in a p-value $<.001$. This result indicates that the null hypothesis can be rejected. Thus, this indicates that the students' scores in the post-test were significantly higher than the pre-test.

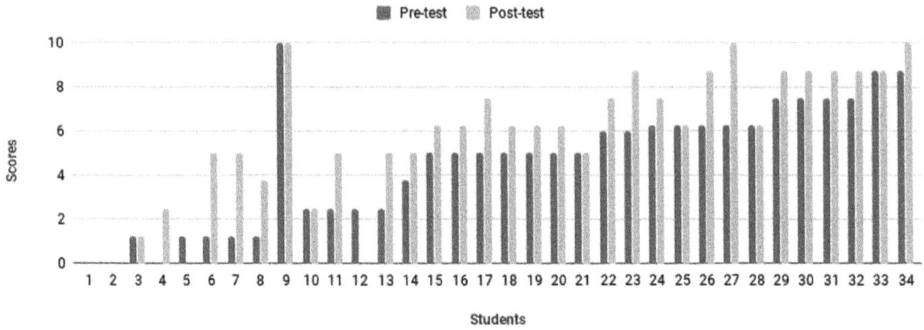

Fig. 3. Results of the pre-test and post-test of the experiment applied with the 34 students.

These results indicate that our feedback approach obtained a positive impact on the students learning. The Table 1 shows the students' requests for feedback during the resolution of the programming questions in the experiment. The most requested type of feedback were the text hints (41.2%).

Table 1. Statistical data of feedback requests made by students during the experiment.

Feedback format	N	%	M	Mdn	SD
Text	80	41.2%	2.35	1.0	3.0
Flowchart	51	36.3%	1.50	0.5	2.2
Video	63	32.5%	1.85	1.0	2.5
Total	194	100%	5.70	1.0	2.6

During the second session of the experiment, students answered three questions with the help of our feedback approach.

We analyze the students' interaction log with the system to check their behaviors during feedback requests. We divided the students into four groups: a group that submitted correct solutions to the three questions, a group that hit two questions, a group that got just one question right, and a group that failed to submit any correct solution. In Fig. 4 a chart is presented with the requests of each feedback format organized according to the four groups created.

During the analysis of the log of interactions, it was noticed that some students of the group that submitted only a correct solution and the group that did not hit any question showed an atypical behavior. They made several requests for feedback in sequence without using incoming messages to change their solution. This behavior has been categorized as help abuse [1]. By removing these students from the analysis, we obtain the feedback request chart shown in Fig. 5.

Regarding the group of students who did not correctly answer any questions during the second session of the quasi-experiment, after analyzing the log of interactions, we noticed that 85% of the feedback messages received by this group were useful for the student to progress in his solution of the problem. In many cases, students have correctly used the feedback messages, but have spent much time between receiving the

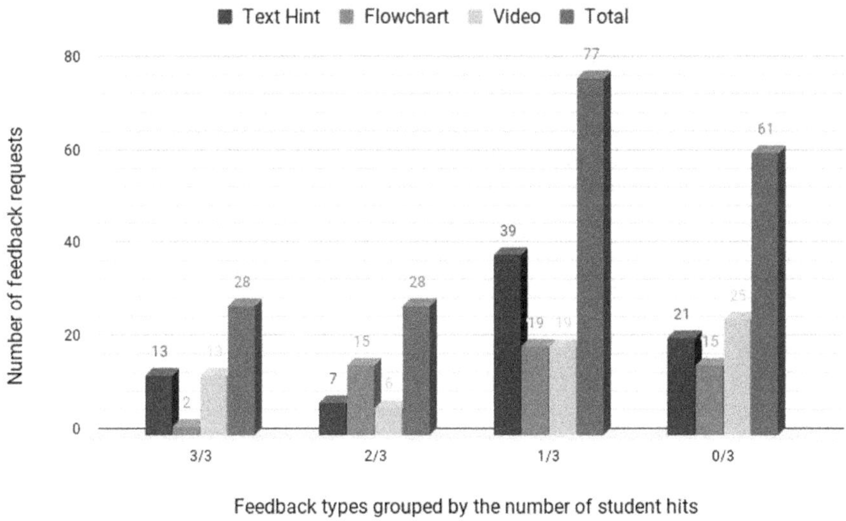

Fig. 4. Chart of feedback requests during the experiment.

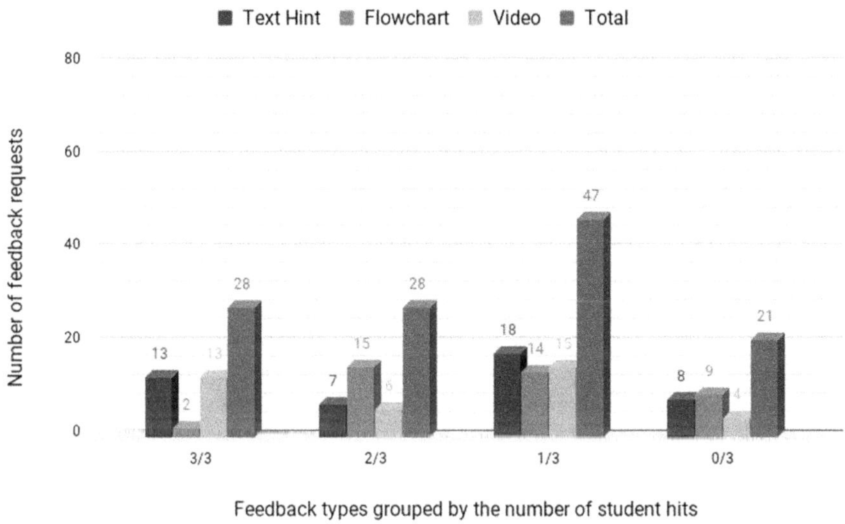

Fig. 5. Feedback requests during the experiment without students with behavior help abuse.

message and applying the hint in their solution. Possibly, if more time were provided in the second session of the quasi-experiment, students could have correctly finalized at least one question.

5.4 Threats to Validity

Although there was concern about the permanence time of the sessions, some students may have felt frustration and boredom during the experiment. Another possible threat to the validity of the results is the representativeness of the sample since all thirty-four students are from the same class and course. This problem can be solved applying this study with different samples and other contexts.

6 Concluding Remarks and Future Perspectives

In this paper, we present an approach to provide adaptive feedback during solving programming problems. The recommendation of feedback is based on the current state of the student's solution. One of the main contribution of this paper is the provision of more than one feedback format (text, video, and flowchart). We found evidence that our approach may have had a positive impact on students' programming learning.

An immediate future work, we have planned to perform an experiment during a full academic semester and studies on the automatic selection of feedback format based on students' learning style. In addition, we intend to compare students' perception and attitudes when using different feedback formats.

References

1. Aleven, V., Koedinger, K.R.: Limitations of student control: do students know when they need help? In: Gauthier, G., Frasson, C., VanLehn, K. (eds.) ITS 2000. LNCS, vol. 1839, pp. 292–303. Springer, Heidelberg (2000). https://doi.org/10.1007/3-540-45108-0_33
2. Cobertt, A.T., Anderson, J.R.: Locus of feedback control in computer-based tutoring: impact on learning rate, achievement and attitudes. In: Proceedings of the SIGCHI Conference on Human Factors in Computing Systems, CHI 2001, pp. 245–252. ACM, New York (2001)
3. Crow, T., Luxton-Reilly, A., Wuensche, B.: Intelligent tutoring systems for programming education: a systematic review. In: Proceedings of the 20th Australasian Computing Education Conference, ACE 2018, pp. 53–62. ACM, New York (2018)
4. Douce, C., Livingstone, D., Orwell, J.: Automatic test-based assessment of programming: a review. J. Educ. Resour. Comput. **5**(3), 4 (2005)
5. Keuning, H., Heeren, B., Jeuring, J.: Strategy-based feedback in a programming tutor. In: Proceedings of the Computer Science Education Research Conference, CSERC 2014, pp. 43–54. ACM, New York (2014)
6. Keuning, H., Jeuring, J., Heeren, B.: A systematic literature review of automated feedback generation for programming exercises. ACM Trans. Comput. Educ. **19**(1), 3:1–3:43 (2018)
7. Kyrilov, A., Noelle, D.C.: Do students need detailed feedback on programming exercises and can automated assessment systems provide it? J. Comput. Sci. Coll. **31**(4), 115–121 (2016)
8. Lahtinen, E., Ala-Mutka, K., Jarvinen, H.M.: A study of the difficulties of novice programmers. In: Proceedings of the 10th Annual SIGCSE Conference on Innovation and Technology in Computer Science Education, ITiCSE 2005, pp. 14–18. ACM, New York (2005)

9. Nguyen, A., Piech, C., Huang, J., Guibas, L.: Codewebs: scalable homework search for massive open online programming courses. In: Proceedings of the 23rd International Conference on World Wide Web, WWW 2014, pp. 491–502. ACM, New York (2014)
10. Rivers, K., Koedinger, K.R.: Data-driven int generation in vast solution spaces: a self-improving python programming tutor. Int. J. Artif. Intell. Educ. **27**(1), 37–64 (2017)
11. Tiam-Lee, T.J., Sumi, K.: Adaptive feedback based on student emotion in a system for programming practice. In: Nkambou, R., Azevedo, R., Vassileva, J. (eds.) ITS 2018. LNCS, vol. 10858, pp. 243–255. Springer, Cham (2018). https://doi.org/10.1007/978-3-319-91464-0_24

Analysis and Prediction of Student Emotions While Doing Programming Exercises

Thomas James Tiam-Lee(✉) and Kaoru Sumi

Future University Hakodate, Hakodate, Japan
g3117002@fun.ac.jp

Abstract. The modeling of student emotions has recently considerable interest in the field of intelligent tutoring systems. However, most approaches are applied in typical interaction models characterized by frequent communication or dialogue between the student and the tutoring model. In this paper, we analyze emotions while students are writing computer programs without any human or agent communication to induce displays of affect. We use a combination of features derived from typing logs, compilation logs, and a video of the students' face while solving coding exercises and determine how they can be used to predict affect. We find that combining pose-based, face-based, and log-based features can train models that predict affect with good accuracy above chance levels and that certain features are discriminative in this task.

Keywords: Student modeling · Affective computing · Programming

1 Introduction and Related Works

Over the past decades, there has been a rapid increase in interest in intelligent tutoring systems (ITS). A valuable component in the design of these systems is the student model, which keeps track of a student's state during the learning process. In recent years, particular interest has been placed on modeling not only the cognitive but also the affective states of students while interacting with tutoring systems [14]. These endeavors are supported by a host of studies that empirically correlate affect with student achievement [8] and self-regulated learning [6, 15].

Much work has been done recently in the recognition of student affect [16]. In these studies, data is collected from students' interactions with the system as well as from various sensors. Machine learning models built from these features are effective in predicting student affect. Features that are commonly used across different studies are facial features (expressions, the location of facial points), eye gaze, posture, and system logs (information about how the student interacted with the system).

However, most of the work on affect detection were applied on systems that utilize typical ITS interaction models wherein students spend time viewing learning materials, answering questions, requesting hints, and communicating with the tutor model. In this kind of interaction model, there is frequent interactivity between the student and the tutor, thus providing plenty of avenues to induce displays of emotion. However, there are other subclasses of ITS that do not feature the same type of interaction.

A. Coy et al. (Eds.): ITS 2019, LNCS 11528, pp. 24–33, 2019.
https://doi.org/10.1007/978-3-030-22244-4_4

A prominent example is the intelligent programming tutor (IPT), a subclass of ITS designed for learning programming.

IPTs pose challenges that are not present in traditional ITSs [7]. In these systems, the majority of the interaction between the student and the system is spent on building programs, which include writing, testing, and debugging code. As a consequence, interactivity between the student and the tutor in these systems are usually fewer and far in between, if not absent. Despite this, previous studies have shown that students frequently transition between different academic emotions while taking part in this kind of activity [4, 5]. Thus, there is much value in being able to infer the affective states of students in this context.

Studies that investigate the detection of academic emotions in the context of an IPT are relatively limited. In a series of studies by Grafsgaard et al. [11–13], posture, gesture, and facial features are used to predict various academic emotions in a programming context. However, the setup involved a human tutor communicating with the student through an interface, a deviation from how typical IPTs operate.

More recently, a study by [3] used facial action coding system features to build models to detect academic emotions while programming. It was found that fixed-point judgments (the students did not decide the intervals) could not be detected in levels above chance, although confusion and frustration could be detected slightly more reliably if only spontaneous judgments (the students decided the intervals) were considered. In our previous work, an IPT that generates programming exercises and offers guides based on the presence of confusion was developed [18].

In this study, we combine head pose, facial expressions, and log-based features in detecting academic emotions on a fine-grained level. We believe that our research can enable the design of better IPTs by allowing them to respond to student emotional states. For example, if confusion is detected, the system may automatically respond by providing more similar patterns of problems. On the other hand, if frustration is detected, the system may react with motivational messages.

2 Data Collection

In this section, we discuss our methodology for data collection. Our objective is to analyze a combination of pose-based, face-based, and log-based features in a coding activity vis-a-vis the academic emotions that were reported by students. To accomplish this, we asked university students enrolled in a first year programming class to take part in a simulated programming session.

Each student participated in the session individually. Each student used a desktop computer with a webcam on top of the monitor at the center to capture the face. The session was divided into two parts: the coding phase and the self-report phase. In the coding phase, the student must solve coding exercises that covered introductory programming concepts. In each exercise, the student must write the body of a function according to given specifications. 9 exercises must be solved in sequential order. It was not allowed to move to the next exercise until a correct solution was submitted. The coding phase lasted for 45 min or until all exercises were solved successfully.

A custom application was used for the coding phase which provided an interface for writing and running code. It also provided basic IDE-like features such as automatic indention and syntax error checking. This application automatically collected and logged the following information throughout the coding phase: all changes in the code (insertions and deletions), all code compilations and submissions, and a video of the student's face. Figure 1 shows the setup of the data collection and a screenshot of the system used.

After the coding phase, the student moved on to the self-report phase. Each student was shown a replay of the session, which could be freely controlled with the use of a slider bar to adjust the time. The system partitions the session data into intervals based on key moments such as: (1) the start and end of a series of key presses, (2) compilation of the code, (3) submission of the code, and (4) the start of a new problem. For each interval, the student must provide one affective state label (engaged, confused, frustrated, bored, neutral) and one action state label (reading, thinking, writing, finding, fixing, unfocused, other). To minimize subjectivity on the judgments, we provided the definitions for each affective and action state label. For the four neutral affective states, we defined engaged as "you are immersed in the activity and enjoying it", confused as "you have feelings of uncertainty on how to proceed", frustrated as "you have strong feelings of anger or disappointment", and bored as "you feel a lack of interest in continuing with the activity".

The average length of the intervals is 17.24 s, resulting in fairly fine-grained and fixed-point affect judgment data. A maximum of 150 intervals for each student was set to keep the annotation task manageable. For sessions that contained more than 150 intervals after partitioning, 150 intervals were randomly selected for the student to annotate. A total of 73 students[1] participated (38 in Japan and 35 in the Philippines), resulting in a combined total of 49 h, 25 min and 17 s of collected session data.

We used OpenFace to extract information from the raw videos of the programming session data. OpenFace is a computer vision toolkit capable of head pose estimation, eye gaze estimation, and action unit recognition using state-of-the-art machine learning approaches [2]. We extracted the following information for each frame of the videos: the position of the head in 3-D space (in mm), the rotation of the head (in radians), the eye gaze angle in world coordinates (in radians), and the intensity of each action unit (a number from 0 to 5). Action units are a taxonomy of fundamental actions of facial muscles used in previous studies for emotion recognition [10]. An example of an action unit is raising the inner brow. OpenFace has good inter-rater agreement with human baselines across multiple datasets in AU detection [1].

In an effort to make the simulated programming sessions natural, we did not restrict the posture and movement of the students during the sessions. Unfortunately, this resulted in some videos where facial landmark estimation resulted in low confidence due to non-ideal positions of the head with respect to the camera throughout the majority of the session. We excluded these videos from analyses involving facial features. After removing these from the data, we were left with 20 videos for the Japanese group and 19 videos for the Filipino group. Furthermore, we only consider

[1] All students gave us the permission to publish their faces in academic publications.

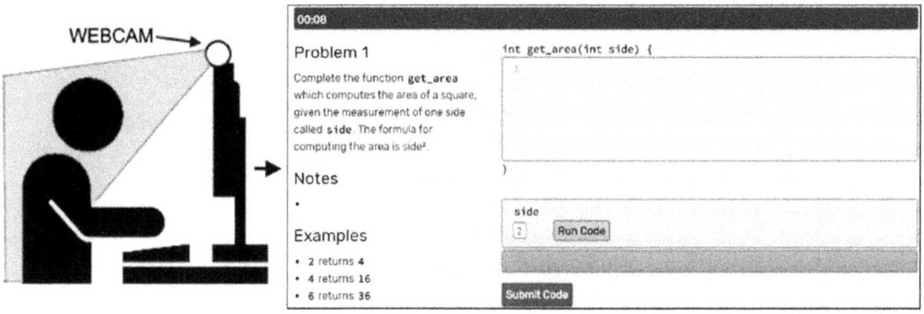

Fig. 1. Setup and system used for programming practice

frames which have high confidence of facial landmark estimation by applying some filtering heuristics to clean the data.

3 Results of Analysis

In this section, we discuss the results of our analysis on the data. First, we discuss models for classifying the reported affective states, and then we discuss individual features that were useful in predicting affect.

3.1 Detecting the Presence of Affective States

We built classifier models using WEKA to determine whether it was possible to recognize naturalistic displays of affect while programming. In these models, we only considered intervals that are at least 10 s in length and in which OpenFace was successfully able to detect the face. 10-fold cross-validation was used to evaluate the models. In each fold, 90% of the intervals were used for training, and 10% of the intervals were used for testing. First, we determined if it was possible to detect the presence or absence of each affective state (i.e., confused or not confused) according to the student reports. We used C4.5 decision trees, support vector machines, multilayer perceptron, and naive Bayes classifier and chose the best performing type of model for each task. The results are shown in Table 1.

Because the classes are imbalanced in these models, the metric we used is Cohen's kappa value, which measures how well the model performs against a model that simply guesses randomly based on the frequency of each class. For both groups, using pose-based and face-based features alone or log-based features alone resulted in models that do not perform much beyond random chance $\kappa < 0.2$ except for engagement. However, by using a combination of these features, the performance of the models could be increased. By using pose-based, face-based, and log-based features together, we are able to come up with models that achieve performance beyond chance for engagement and frustration, but confusion and boredom remained to be difficult to detect. For confusion, we were able to improve the performance of the model to 0.75 accuracy ($\kappa = 0.22$) for the Japanese group and 0.72 accuracy ($\kappa = 0.2$) for the Filipino group by

Table 1. Accuracy and Cohen's Kappa (in parenthesis) for classifying the presence and absence of affective states

Japanese group				
Features	Engaged	Confused	Frustrated	Bored
pose + face	0.69 (0.39)	0.69 (0.11)	0.73 (0.13)	0.90 (0.13)
log	0.63 (0.26)	0.77 (0.01)	0.85 (0.07)	0.93 (0.00)
pose + face + log	0.71 (0.42)	0.71 (0.18)	0.82 (0.27)	0.91 (0.16)
pose + face + log + action	0.76 (0.51)	0.72 (0.16)	0.82 (0.32)	0.93 (0.39)
Filipino group				
Features	Engaged	Confused	Frustrated	Bored
pose + face	0.65 (0.28)	0.67 (0.08)	0.71 (0.16)	0.93 (0.17)
log	0.58 (0.07)	0.75 (0.00)	0.75 (0.00)	0.95 (0.00)
pose + face + log	0.65 (0.30)	0.67 (0.10)	0.72 (0.22)	0.93 (0.10)
pose + face + log + action	0.66 (0.31)	0.69 (0.14)	0.73 (0.24)	0.95 (0.39)

filtering the types of pose-based and face-based features used using RELIEF-F feature selection. Adding the reported action state as a feature further improved the performance of the models significantly. However, it should be noted that the reported action state is not an observable feature unlike the others.

3.2 Meaningful Features in State Detection

In order to investigate which features are useful in predicting emotion, we used RELIEF-F feature ranking to determine the important features in detecting the presence of affective states. RELIEF-F ranks features based on their ability to discriminate an instance from its closest neighbors that belong to a different class. In this section we discuss these features.

AU04 - Brow Lowerer and Other Eye-Related AU. The AU04 action unit, which pertains to the "brow lowerer" action was ranked as an important feature in all of the classification tasks. This AU was also found in previous studies to be associated with the negative states of confusion and frustration [3, 11]. Figure 2 shows some of the prominent displays of AU04. This AU was observed in the data when the eyebrow is furrowed, as well as when the subjects gaze downwards. This often happened during typing when the eyes quickly shift between looking at the screen and at the keyboard without turning the head down. The mean intensity display of AU04 is higher in reported intervals of frustration than during intervals of engagement or confusion, as shown in Table 2. Furthermore, by performing a Wilcoxon signed ranked test across all the subject data, we found that the mean intensity of AU04 was significantly higher in reported intervals of frustrations than in all the other states combined. This is summarized in Fig. 3a.

Lip-Related AUs. Although there is no significant correlation between AU12 and AU10 and reported affective states, a Wilcoxon signed ranked test revealed that AU10 was displayed in significantly higher intensities during reported intervals of reading

Fig. 2. From Left to Right (in pairs): AU04 (brow lowerer), AU12 (Lip corner puller), AU10 (Upper lip raiser)

Table 2. Mean intensity display of AU04 in Typing and Non-typing intervals

	Typing		Non-typing	
	Japanese	Filipino	Japanese	Filipino
Engaged	0.64	0.21	0.89	0.21
Confused	0.66	0.22	0.65	0.29
Frustrated	**0.98**	**0.31**	**1.35**	**0.35**

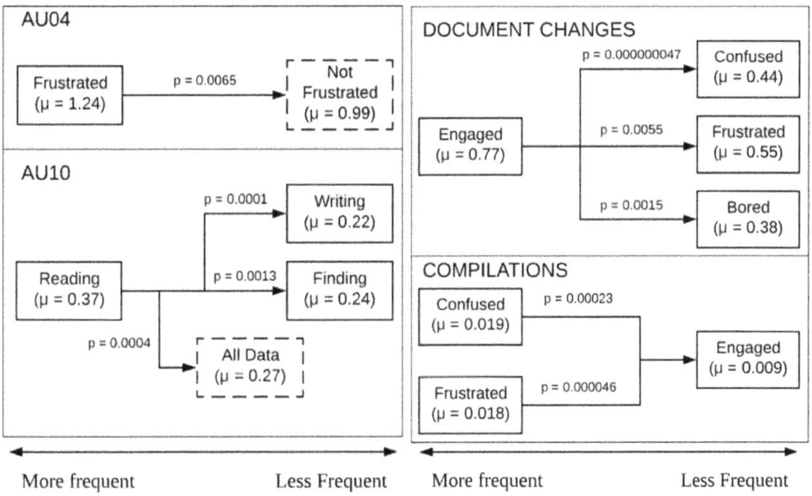

Fig. 3. Correlation between (a) AU and states, and (b) logs and states. Arrows point from a state that occurs significantly greater than the one pointed to.

than in reported intervals of writing and finding, and all the other states combined. This is summarized in Fig. 3a. Examples of displays of these AUs are shown in Fig. 2.

Head Location. One feature that was ranked highly in models classifying boredom was the head location standard deviation features. An increase in the mean standard deviation of all head location features can be observed in reported intervals of boredom as compared to the other affective states, as shown in Table 3. This suggests that there tends to be a wider range of head movement.

Table 3. Head location average standard deviation across different emotions

	Japanese group				Filipino group			
	Eng.	Conf.	Frus.	Bored	Eng.	Conf.	Frus.	Bored
Loc. X	12.10	15.75	15.00	**20.87**	14.14	14.06	14.34	**17.25**
Loc. Y	10.16	10.55	11.58	**15.97**	9.80	9.02	9.01	**10.68**
Loc. Z	13.91	15.67	16.67	**23.27**	10.75	11.73	11.96	**15.74**

Document Insertions and Code Compilations. Performing a paired Wilcoxon signed rank test on each combination of feature and affective state with a Bonferonni correction to account for 6 hypotheses resulting in $\alpha = 0.0083$, we found that document insertions occurred significantly more when students were engaged ($\mu = 0.65$) than when they were confused ($\mu = 0.34$, $p = 0.000000011$), frustrated ($\mu = 0.35$, $p = 0.000046$) or bored ($\mu = 0.25$, $p = 0.00078$). When document insertions and deletions are taken together, we found that document changes occurred significantly more when students are engaged ($\mu = 0.77$) then when they are confused ($\mu = 0.44$, $p = 0.000000047$), frustrated ($\mu = 0.55$, $p = 0.0055$) or bored ($\mu = 0.38$, $p = 0.0015$). On the other hand, we found that compilations with syntax errors occurred significantly less when students were engaged ($\mu = 0.0037$) than when they were confused ($\mu = 0.0086$, $p = 0.0005$) or frustrated ($\mu = 0.0089$, $p = 0.0044$). When all code compilations were considered, we similarly found that significantly less compilations occurred when students were engaged ($\mu = 0.0093$) than when they were confused ($\mu = 0.019$, $p = 0.000023$) or frustrated ($\mu = 0.018$, $p = 0.000046$). In addition, we also found that compilations occurred significantly more when students were frustrated than when they were bored ($\mu = 0.015$, $p = 0.0043$). The findings are summarized in Fig. 3b. Overall, more document insertions were indicative of engagement, while more code compilations were indicative of confusion and frustration.

Student's Reported Action State. We took each interval and for each emotion, we calculated the probability it occurred with each action state, then we performed paired Wilcoxon signed ranked tests with a Bonferonni correction to account for 15 hypotheses resulting in $\alpha = 0.0033$. Figure 4 shows the results. We found that engagement was reported the most while writing, while confusion and frustration were reported the most while thinking. Finally, boredom was reported significantly more while unfocused, writing, or thinking than while reading or fixing bugs.

4 Discussion

While there has been a lot of work in detection of emotion in intelligent tutoring systems, naturalistic displays of affect such as that in intelligent programming tutors are relatively more difficult because the emotions are not directly induced by human or agent interactions [9]. A previous study has shown the fixed point judgments (affect report intervals are not decided by the student) are difficult to detect using face and posed-based features as compared to spontaneous judgments (affect report intervals are decided by the student). This finding was supported by our data which also used fixed

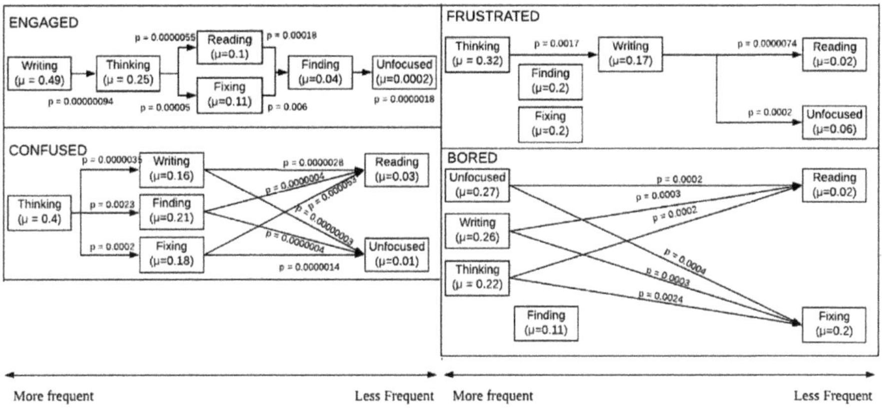

Fig. 4. Occurrence of action states and emotion states. The arrows point from an action that occurs significantly greater than the action being pointed to.

point judgments, as confusion, frustration, and boredom could not reliably be classified beyond κ > 0.2. However, combining face-based and pose-based features with log-based features can improve the performance of classifiers in detecting fixed point affect judgments. Using these combinations of features, we can achieve classifiers that could perform well beyond random chance.

In our analysis, we also found that adding the action state reported by the student can increase the performance of the models. This was also supported by the action state appearing prominently as a highly-ranked feature in most training iterations, as well as observations that students are likely to report certain affective states when in certain action states. This shows that emotion of students while programming could be better inferred with a good understanding of the student's behavior. The problem with this is that the reported action state is not an observable feature like the others, and can be thought of as another output of a stochastic process. Nevertheless, this suggests that modeling student's subtasks and intentions in programming (finding the bug, writing the code, etc.) may help improve the performance of classifying affect.

Another finding in this study is that log-based, face-based and pose-based features are all meaningful in predicting affect. We found evidence supporting previous studies linking AU04 to confusion and frustration, but we also found other expressions such as AU10 and AU12 that, despite not being correlated to affect, were discriminative and were more closely tied to actions. We also found that the standard deviation of the head location can also be indicative of boredom. Log-based features such as the number of document insertions and compilations are also meaningful in prediction.

In this study, we did not find any significant differences between the Japanese and Filipino groups, aside from minor variations on the ranking of important features. The total of the mean intensities across all AUs was higher among Japanese students compared to Filipino students, suggesting that in this study the Japanese students showed more facial expressions. However, this contradicts findings from a previous study in which the Filipino students had stronger displays of facial expressions [17]. The small sample size in these studies is insufficient to generalize on the populations.

Affect detection is generally considered to be a challenging task caused by a variety of factors such as noise, individual differences, and the general complexities of human behavior. Although this study has limitations including reliance on OpenFace's facial landmark and AU detection, non-consideration of co-occurring affective states, and a question of generalizability on other datasets, we were able to show that predicting student emotions while interacting with a system that does not directly induce displays of affect can be achieved by using a combination of observable physical features as well as log information.

5 Conclusion

In this paper, we used a combination of pose-based, face-based, and log-based features to model naturalistic displays of student affect while writing computer programs. By using a combination of those features, we were able to train models that perform above chance levels. We also presented features that were useful for the detection of affect in programming. For the future direction of this study, we intend to explore further additional features that can be used for emotion detection, and explore how they could be applied in ITS for teaching students how to code.

Acknowledgments. The authors would like to thank Mr. Fritz Flores, Mr. Manuel Toleran, and Mr. Kayle Tiu for assisting in the facilitation of the data collection process in the Philippines.

References

1. Baltrusaitis, T., Mahmoud, M., Robinson, P.: Cross-dataset learning and person-specific normalisation for automatic action unit detection. In: 2015 11th IEEE International Conference and Workshops on Automatic Face and Gesture Recognition (FG), vol. 6, pp. 1–6. IEEE (2015)
2. Baltrusaitis, T., Zadeh, A., Lim, Y.C., Morency, L.P.: Openface 2.0: facial behavior analysis toolkit. In: 2018 13th IEEE International Conference on Automatic Face & Gesture Recognition (FG 2018), pp. 59–66. IEEE (2018)
3. Bosch, N., Chen, Y., D'Mello, S.: It's written on your face: detecting affective states from facial expressions while learning computer programming. In: Trausan-Matu, S., Boyer, K.E., Crosby, M., Panourgia, K. (eds.) ITS 2014. LNCS, vol. 8474, pp. 39–44. Springer, Cham (2014). https://doi.org/10.1007/978-3-319-07221-0_5
4. Bosch, N., D'Mello, S.: Sequential patterns of affective states of novice programmers. In: The First Workshop on AI-supported Education for Computer Science (AIEDCS 2013), pp. 1–10 (2013)
5. Bosch, N., D'Mello, S., Mills, C.: What emotions do novices experience during their first computer programming learning session? In: Lane, H.C., Yacef, K., Mostow, J., Pavlik, P. (eds.) AIED 2013. LNCS (LNAI), vol. 7926, pp. 11–20. Springer, Heidelberg (2013). https://doi.org/10.1007/978-3-642-39112-5_2
6. Cho, M.H., Heron, M.L.: Self-regulated learning: the role of motivation, emotion, and use of learning strategies in students learning experiences in a self-paced online mathematics course. Distance Educ. **36**(1), 80–99 (2015)

7. Crow, T., Luxton-Reilly, A., Wuensche, B.: Intelligent tutoring systems for programming education: a systematic review. In: Proceedings of the 20th Australasian Computing Education Conference, pp. 53–62. ACM (2018)
8. Daniels, L.M., Stupnisky, R.H., Pekrun, R., Haynes, T.L., Perry, R.P., Newall, N.E.: A longitudinal analysis of achievement goals: from affective antecedents to emotional effects and achievement outcomes. J. Educ. Psychol. **101**(4), 948 (2009)
9. D'Mello, S., Kory, J.: Consistent but modest: a meta-analysis on unimodal and multimodal affect detection accuracies from 30 studies. In: Proceedings of the 14th ACM International Conference on Multimodal Interaction, pp. 31–38. ACM (2012)
10. Ekman, P., Friesen, W.V.: Unmasking the Face: A Guide to Recognizing Emotions from Facial Cues (1975)
11. Grafsgaard, J.F., Boyer, K.E., Lester, J.C.: Predicting facial indicators of confusion with hidden Markov models. In: D'Mello, S., Graesser, A., Schuller, B., Martin, J.-C. (eds.) ACII 2011. LNCS, vol. 6974, pp. 97–106. Springer, Heidelberg (2011). https://doi.org/10.1007/978-3-642-24600-5_13
12. Grafsgaard, J.F., Wiggins, J.B., Boyer, K.E., Wiebe, E.N., Lester, J.C.: Automatically recognizing facial indicators of frustration: a learning-centric analysis. In: 2013 Humaine Association Conference on Affective Computing and Intelligent Interaction, pp. 159–165. IEEE (2013)
13. Grafsgaard, J.F., Wiggins, J.B., Boyer, K.E., Wiebe, E.N., Lester, J.C.: Embodied affect in tutorial dialogue: student gesture and posture. In: Lane, H.C., Yacef, K., Mostow, J., Pavlik, P. (eds.) AIED 2013. LNCS (LNAI), vol. 7926, pp. 1–10. Springer, Heidelberg (2013). https://doi.org/10.1007/978-3-642-39112-5_1
14. Harley, J.M., Lajoie, S.P., Frasson, C., Hall, N.C.: Developing emotion-aware, advanced learning technologies: a taxonomy of approaches and features. Int. J. Artif. Intell. Educ. **27** (2), 268–297 (2017)
15. Mega, C., Ronconi, L., De Beni, R.: What makes a good student? How emotions, self-regulated learning, and motivation contribute to academic achievement. J. Educ. Psychol. **106**(1), 121 (2014)
16. Petrovica, S., Anohina-Naumeca, A., Ekenel, H.K.: Emotion recognition in affective tutoring systems: collection of ground-truth data. Procedia Comput. Sci. **104**, 437–444 (2017)
17. Tiam-Lee, T.J., Sumi, K.: A comparison of Filipino and Japanese facial expressions and hand gestures in relation to affective states in programming sessions. In: Workshop on Computation: Theory and Practice 2017 (2017)
18. Tiam-Lee, T.J., Sumi, K.: Adaptive feedback based on student emotion in a system for programming practice. In: Nkambou, R., Azevedo, R., Vassileva, J. (eds.) ITS 2018. LNCS, vol. 10858, pp. 243–255. Springer, Cham (2018). https://doi.org/10.1007/978-3-319-91464-0_24

Analyzing the Group Formation Process in Intelligent Tutoring Systems

Aarón Rubio-Fernández(✉), Pedro J. Muñoz-Merino, and Carlos Delgado Kloos

Universidad Carlos III de Madrid, Leganés, Spain
{aarubiof, pedmume, cdk}@it.uc3m.es

Abstract. Due to the increasing use of active learning methodologies such as the Flipped Classroom, which usually use group activities, Intelligent Tutoring Systems (ITSs) have the opportunity of supporting them. In this work, we present a group formation tool that allows teachers to create easily and efficiently groups of students based on the K-means algorithm. Our tool allows creating the groups using videos' and exercises' information. Moreover, we have validated the suitability of our group formation process and tool through the students' scores. In order to gain insights into the group formation in learning systems which only use exercises or use mainly videos, we also compare the groups through the Jaccard similarity measurement. Results show that the groups formed using only the videos' information are quite similar to the groups created using just the exercises' information. Therefore systems that only use exercise interactions for group formation might be replaced by others that only use video interactions and the other way around.

Keywords: Group formation · Flipped classroom · Intelligent tutoring systems · Collaborative learning

1 Introduction

The increasing use of active learning methodologies such as the Flipped Classroom (FC) has to be supported by Intelligent Tutoring Systems, in order to provide learners with a suitable learning context. In a FC, learners access to academic resources (e.g. videos or exercises) before the face-to-face lessons in order to prepare them. During the lesson, teachers try to promote learners' engagement using active learning activities, like group projects [1] that promote collaborative learning and interdisciplinary skills, which can lead to successful learning experiences [2].

Exercises and videos are the most predominant resources in the FC, and they are also used in ITSs and other educational contexts such as MOOCs (Massive Open Online Course). For this reason, we focus on the data related to these resources. Furthermore, we propose a tool to form groups based on the use of the previous types of data, to reduce the teachers' workload related to the Group Formation Process (GFP). This workload reduction can promote the use of group activities in many contexts such as the ones which use the FC [3], since teachers are allowed to focus only on the design of the group activity instead of on the GFP.

A. Coy et al. (Eds.): ITS 2019, LNCS 11528, pp. 34–39, 2019.
https://doi.org/10.1007/978-3-030-22244-4_5

Furthermore, we try to answer the following research questions: Are the groups created through our tool suitable for group activities? Are the formed groups the same using different type of data (e.g. only information about exercises or videos)? Would the GFP be the same e.g. in an ITS driven by exercises in comparison with a MOOC driven just by videos?

2 Related Work

The GFP has associated different approaches in terms of the data used. For example, it is possible to use the data gathered through tests [4]; the learners' grades [5]; or other alternatives shown in the literature which depend on the learning context and the purposes of the study. In our work, the data are related to the videos and exercises used to prepare the face-to-face sessions of a FC.

We are interested to know if the formed groups are really homogeneous or heterogeneous, and whether the decisions for the group formation would be the same based on videos or exercises (both are the most used resources in the FC), since, as far as we know, this type of analysis has not been done. If decisions are the same, then the decisions based on exercises (which are mainly used in ITSs) would be equivalent to the ones based on videos (which are e.g. predominant in MOOCs).

In the case of the group formation, there are two different approaches which can be used to cluster (classify) learners: (1) approaches focused on the use of low complexity algorithms, and (2) approaches that cluster learners using more complex algorithms. In the first case, the most used low complexity algorithms are the K-means algorithm [6] and the Fuzzy C-Means [7]. Regarding more complex algorithms, some possible examples are the "Expectation-maximization" algorithm [8]; or optimization algorithms such as the "Ant Colony Optimization" or the "Genetic Algorithms" [9].

3 Group Formation Algorithm and Tool

Many learning systems provide students with educational videos and exercises, but, in some cases, there is only available information associated either the exercises or the videos. For this reason, we propose two types of data to be used in the GFP: (1) videos' information; and (2) exercises' information. In order to create the groups, we use these data and the K-means algorithm [10] which allows assigning the learners into different groups (clusters) based on common characteristics (profiles). Using the learner profiles (discovered through a training data set of 3543 learners enrolled in 19 engineering courses), we classify new learners according to their profiles, and this allows us to create the groups. If the groups are homogeneous, we assign students with a common profile to each group; while if the groups are heterogeneous, each group contains students with different profiles.

Regarding the GFP, we follow a similar approach for all of the group formation methods proposed in this work. We use a set of vectors that contain the learners' data as an input of the K-means algorithm, and this algorithm obtains the learner profiles used to create the groups. The differences among the approaches are related to the input vectors which are different in each case. If we create the groups using only the videos' information, the vectors represent the number of times that each second of the videos has been watched. To create the groups with only the exercises' information, we use the number of attempts, and an indicator defined as the division between the "number of times that the exercise has been solved correctly" and the "number of times that the exercise has been attempted". This indicator allows characterizing behaviors like when a student tries many times an exercise, but they are not be able to solve it. Finally, to create the groups using the videos' information and the exercises' information, we aggregate both types of information in one vector.

3.1 The Group Formation Tool

Our group formation tool is integrated within an e-learning platform named GEL, which provides teachers and students with an educational environment suitable for the FC model [11]. We have developed the group formation tool through the PHP and Python programming languages. Furthermore, we include the possibility of forming random groups, in order to provide teachers with this type of group formation without using any previous data. An example of the tool's teacher dashboard is shown in Fig. 1.

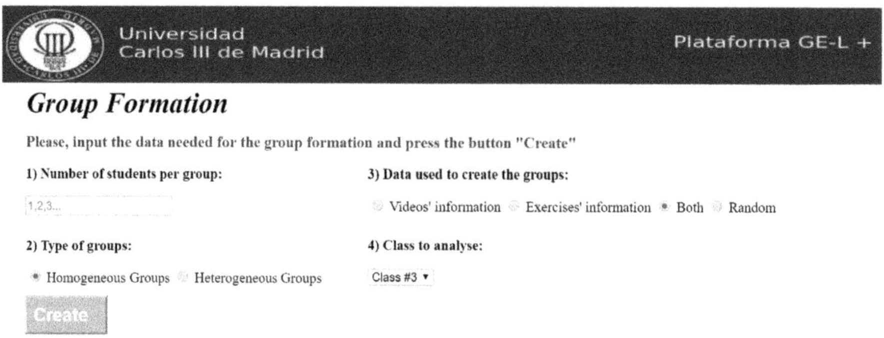

Fig. 1. Tool's teacher dashboard

4 Analysis of the Group Formation

In order to answer our research questions, we have to analyze two aspects: (1) the similarity between groups, and (2) the quality of our GFP. Regarding the student data set used for the analysis, we have gathered data from 162 learners enrolled in five different classes of the same engineering course which uses the FC as learning methodology. The data is related to the whole semester, which in our case lasts 14 weeks involving two sessions of two hours per week (in total four hours each week). These data are different from the training data set used to discover the learner profiles.

4.1 Analyzing the Quality of Our Group Assignation

We analyze the homogeneity and heterogeneity of the groups using the students' scores. The evaluation shows that the GFP is done properly, since the homogeneous groups include students with similar characteristics, while the heterogeneous groups contain students with different characteristics. For example, we obtain the following results in our evaluation (Table 1 shows the scores' statistics of two homogenous groups, while Table 2 presents the scores' statistics related to a couple of heterogeneous groups; the minimum possible score is 0, and the maximum possible score is 10).

Table 1. Homogeneous groups' scores

Group	Mean	Standard deviation
A	6.8	0.4243
B	9.55	0.0707

Table 2. Heterogeneous groups' scores

Group	Mean	Standard deviation
C	6.5	4.2684
D	5.95	4.2852

This example shows the trends of the evaluation of the group formation's quality. Students within a homogeneous group are more similar than those belonging to a heterogeneous group, since the scores of the homogeneous groups have associated standard deviations lower than the ones for the heterogeneous groups. Apart from that, the scores' means of the homogenous groups are more different than the means of the heterogeneous groups; so that students belonging to distinct homogeneous groups are more different, in comparison with the case of the heterogeneous groups where we have similar score patterns in the groups. Therefore, we can infer that the group formation is done properly, since the homogeneous groups and the heterogeneous groups contain students with the appropriate profiles.

4.2 Comparing the Group Formation

In this case, we compare groups created with just videos' information, with groups formed using just exercises' information. For each one of the five analyzed classes, we form groups using each one of these types of information to obtain two group sets. After that, we compare each possible pair of sets using the Jaccard similarity [12], in order to analyze the effects of the academic resource on the GFP. We show the definition of the Jaccard similarity below (Eq. 1).

$$\text{Similarity of two sets} = \frac{a}{a+2\cdot b}\begin{cases} a = \text{number of elements shared by the two sets} \\ b = \text{number of elements not shared by the two sets} \end{cases} \tag{1}$$

The possible values of this indicator of similarity are included in the interval [0, 1]; where the value "0" means that the two groups are totally different (i.e. they do not share any element), and the value "1" implies that both groups are the same (i.e. they share the same elements).

Regarding the similarity between the group sets, we calculate the similarity of each possible combination "group of the first set" - "group of the second set". After that, we sum all the similarity values and divide this aggregated value by the number of groups of the sets (which is the same for the two sets). In this way, we obtain a mean value of similarity, with which compare the groups created using the information related to the exercises with the groups based on the videos' information. In particular, the similarity between the two sets of groups is shown below (Table 3).

Table 3. Similarity between the exercises' information set and the videos' information set

	Class #1	Class #2	Class #3	Class #4	Class #5	Mean	Standard deviation
Heterogeneous groups	0.6314	0.6023	0.6492	0.5942	0.6987	0.634	0,0418
Homogenous groups	0.5967	0.6222	0.5783	0.6318	0.8815	0.6539	0,1244

We can see that the groups created using the two types of information have a similarity below 0.66 in both cases. According to the literature [13], we consider that a value above 0.55 implies that the two sets are similar. Therefore, the two types of group formation are quite similar, since the mean similarities between them (0.634 and 0.6539) are above this threshold (0.55). This implies that the information related to the videos produces a set of groups considerably similar to the groups formed by the information associated with the exercises, so that group formation can be applied in different learning systems with diverse capabilities and academic resources.

5 Conclusions and Future Work

In this work, we analyze the GFP based on information related to videos and on information associated with exercises. Moreover, a tool has been developed to form homogeneous and heterogeneous groups using different criteria: using separately each type of data, taking into account both of them, or using a random assignation. In this way, ITSs which only use exercises or only use videos can form the student groups through our tool.

Furthermore, we measure the quality of our GFP using real student data, and we analyze the similarity between groups created through different academic resources. We use the students' scores in the evaluation of the quality of the created groups, and we analyze the similarity between groups using the Jaccard similarity. The results of these two analyses are the following: (1) the quality of the group formation is adequate; and (2) groups formed using only the videos' information are quite similar to the groups created with only the exercises' information.

However, there are several issues that must be considered. First, the analysis of the group formation can be biased due to the relatively low number of students, so that a broader study with more students is needed in order to generalize the results. Moreover, other algorithms could be used to form the groups, in order to compare them to the groups created by the K-means algorithm. Apart from that, we have to analyze the learning outcomes (e.g. students' scores or students' opinion surveys) obtained in group activities based on the groups formed using the GFP presented in this work.

Acknowledgement. This work has been partially funded by: FEDER/Ministerio de Ciencia, Innovación y Universidades – Agencia Estatal de Investigación/Smartlet project (TIN2017-85179-C3-1-R). In addition, this work has been partially funded by the e-Madrid-CM project with grant no. P2018/TCS-4307, which is funded by the Madrid Regional Government (Comunidad de Madrid), by the Fondo Social Europeo (FSE) and by the Fondo Europeo de Desarrollo Regional (FEDER); This work has also been supported by the RESET project (TIN2014-53199-C3-1-R) funded by the Ministry of Economy and Competitiveness.

References

1. Bergmann, J., Sams, A.: Flip Your Classroom: Reach Every Student in Every Class Every Day. International Society for Technology in Education, Washington (2012)
2. Muñoz-Merino, P.J., et al.: Flipping the classroom to improve learning with MOOCs technology. Comput. Appl. Eng. Educ. **25**(1), 15–25 (2017)
3. Rubio-Fernández, A., Muñoz-Merino, P.J., Delgado Kloos, C.: Scenarios for the application of learning analytics and the flipped classroom. In: 2018 IEEE Global Engineering Education Conference (EDUCON), pp. 1619–1628. IEEE (2018)
4. Moreno, J., Ovalle, D.A., Vicari, R.M.: A genetic algorithm approach for group formation in collaborative learning considering multiple student characteristics. Comput. Educ. **58**(1), 560–569 (2012)
5. van der Laan Smith, J., Spindle, R.M.: The impact of group formation in a cooperative learning environment. J. Account. Educ. **25**(4), 153–167 (2007)
6. Pardos, Z.A., Trivedi, S., Heffernan, N.T., Sárközy, G.N.: Clustered knowledge tracing. In: International Conference on Intelligent Tutoring Systems, pp. 405–410 (2012)
7. Christodoulopoulos, C.E., Papanikolaou, K.: Investigation of group formation using low complexity algorithms. In: Proceedings of PING Workshop, pp. 57–60 (2007)
8. Min, W., et al.: DeepStealth: leveraging deep learning models for stealth assessment in game-based learning environments. In: International Conference on Artificial Intelligence in Education, pp. 277–286 (2015)
9. Gogoulou, A., et al.: Forming homogeneous, heterogeneous and mixed groups of learners. In: 11th International Conference on User Modeling Proceedings of Workshop on Personalisation in E-Learning Environments at Individual and Group Level, pp. 33–40 (2007)
10. Hartigan, J.A.: Clustering Algorithms. Wiley, New York (1975)
11. Argudo, F. C.: Flip-App o cómo incorporar gamificación a asignaturas "Flipped Classroom" basado en la plataforma Open edX. In: EMOOCs 2017, Spanish Track, pp. 25–34 (2017)
12. Lesot, M.J., Rifqi, M., Benhadda, H.: Similarity measures for binary and numerical data: a survey. Int. J. Knowl. Eng. Soft Data Paradig. **1**(1), 63–84 (2009)
13. Niwattanakul, S., Singthongchai, J., Naenudorn, E., Wanapu, S.: Using of Jaccard coefficient for keywords similarity. Proc. Int. MultiConf. Eng. Comput. Sci. **1**(6), 380–384 (2013)

Analyzing the Usage of the Classical ITS Software Architecture and Refining It

Nikolaj Troels Graf von Malotky(✉) and Alke Martens

University of Rostock, 18059 Rostock, Germany
nikolaj.graf_von_malotky@uni-rostock.de

Abstract. The classical ITS software architecture with four components is an old, popular architecture. We analyzed 93 ITS software architectures found in papers with regard to the classical ITS architecture to review the usage of its components, its naming of the components and the addition of new components. In the analysis we detected 13 recurring problems in these architectures. To fix this we propose two refinements to the classical ITS software architecture that should reduce the occurrences of these problems.

Keywords: ITS · Software architecture · Evaluation

1 Introduction

Software engineering teaches us that complex systems, such as ITSs should be designed with a software architecture. We made a statistic of the number of released systems going in the direction of ITSs to summarize the raw number of available, ready to use systems with scientific papers and found what can be seen in Fig. 1. Then we created the same statistic for all the papers which explained a software architecture for ITSs as seen in Fig. 2.

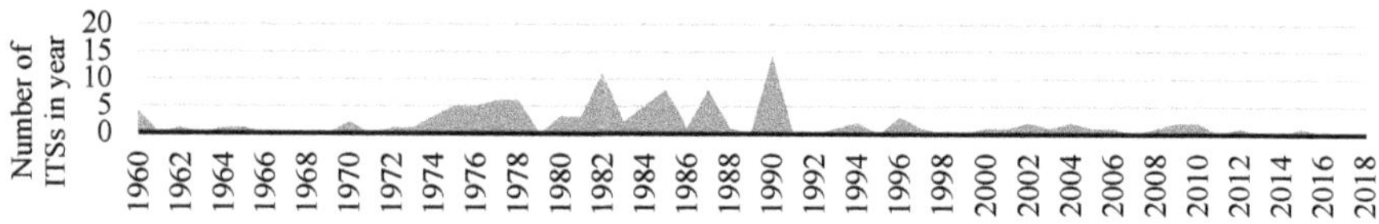

Fig. 1. Number of found ITSs for every year

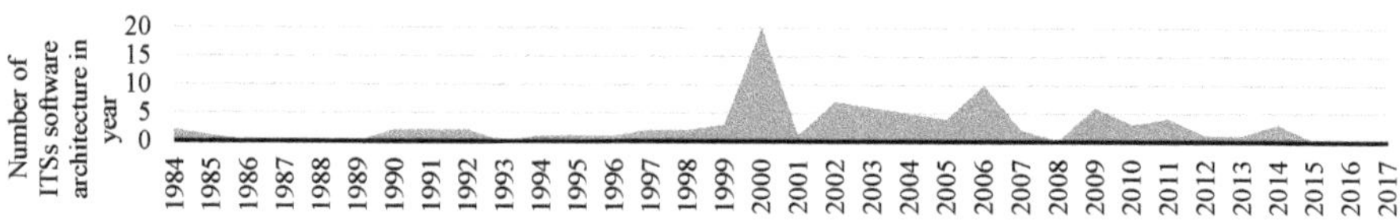

Fig. 2. Number of found ITS software architectures for every year

The current general ITS software architecture is the most used architecture and has not changed much over the last few decades. It is build upon four components (classical approach) and is not uniform. A visually accurate version of the classical architecture is shown in Fig. 3. The general idea is that an ITS should have an interface to

A. Coy et al. (Eds.): ITS 2019, LNCS 11528, pp. 40–46, 2019.
https://doi.org/10.1007/978-3-030-22244-4_6

communicate with students and three components, each having their own category of knowledge. These three knowledge components are: What should be learned by the user, the progress of the user and the behavior for different teaching situations.

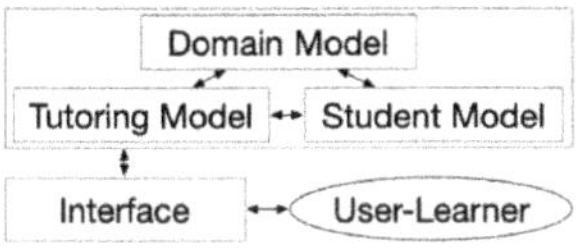

Fig. 3. Classical ITS software architecture [1]

2 Statistics of ITS Software Architectures

We counted how many of the classical four components are represented in the found architectures to show adoption (Fig. 4) and the adoption of every component itself (Fig. 5). A majority of 54%, did cover all classical components, and 82% covered 3 or more components. Since architectures reach back to the beginning of ITS in the 80 s, some of them do not include newer components. The domain model as the oldest component is most often implemented with 86 percent, since the first ITS architecture included an "Expert System" [2].

These components are not usable in a general architecture, since their components are dependent on a specific implementation. As you can see in Fig. 6, half of the architectures are not reusable in a broader area than in the intended application purpose.

20.7% of the architectures had new components in them, which were not implementation specific. Additionally, 14.1% of the architectures did respect an author as a user.

Fig. 4. Percentage of the classical component in the found ITS software architectures

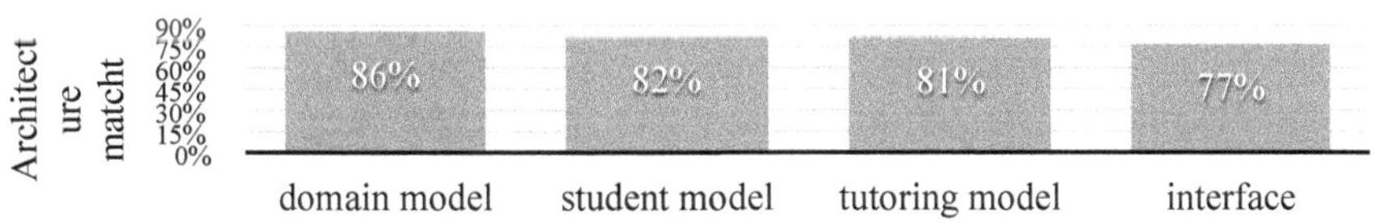

Fig. 5. Percentage of the specific classical component in the found architectures

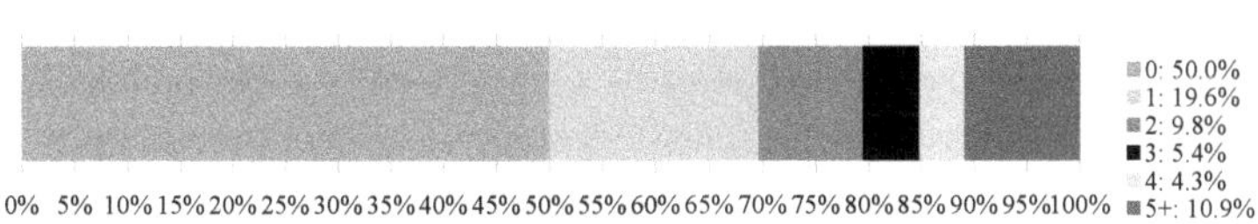

Fig. 6. Number of implementation specific components in the found ITS software architectures

We looked at the number of names for each of the classical components. Hyphens, abbreviations and capitalization were not considered, see Fig. 7. The tutoring model does have the highest count of unique names with 66 different names, followed by the domain model with 47 and the interface with respectively 48 unique names. The student model has the most stable naming with only 41 names. Every component was found to have many names and the naming is not consistent even in current articles.

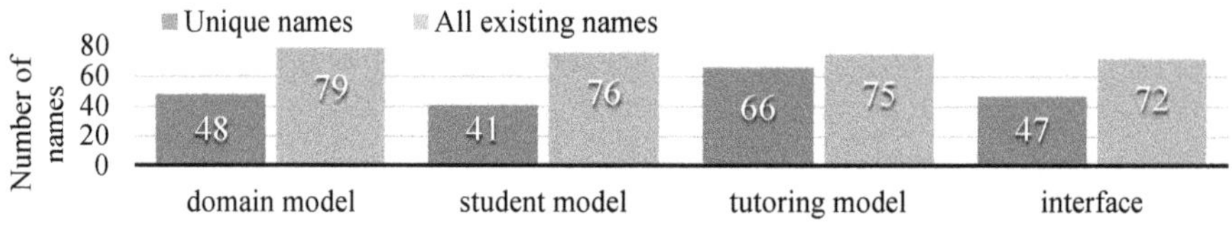

Fig. 7. Number of different component names

3 Problems of the Current Software Architectures

Most of the architectures seem like an afterthought of an already concrete existing ITS or a simple overgeneralization of existing architectures without clear intention of the target group. The different found problems are listed below.

Problem 1: Naming of the classical components is confusing. As seen in Fig. 7 the naming of the components is very different. The architecture from Fig. 3 has well named suffixes for its component, but the naming is additionally confusing because the ending of the components in one architecture are mostly not consistent even when the abstract differences are minimal. It is complicated to remember the names just as aliases for the same functionality and often will result in guesswork and interpretation errors from people reading it.

Problem 2: Changing of the definition of the classical components. Even if the naming of the components is consistent and clear, some architectures did use an already existing name and changed its definition. Only in very rare cases do the architectures explain the source of their definitions and their differences, when using already existing component names. This is especially problematic if the components are classical ones with already established definitions. The reader of the architecture could simply assume that the definition is the already known one. Taking the same name for the new component is very hard to track, which could in the end cause hard to fix problems with the architecture design. Since the ITS development includes so many different parties it is so much more important that all developing members have a clear understanding of the components and their capabilities, responsibilities and limits.

Problem 3: Reinventing an already existing component definition. Some architectures invent components which are already well defined in other architectures. Some found architectures reinvented the classic components without changes and, even worse, the naming is also changed in rare cases. Mostly classical components are split up into multiple parts. Just because a function is very important, you should not separate it from the enclosing component, but describe a classic component more detailed. An error which seems to be made, because of the focus on the new technical implementation of one part of a component. Vice versa the combination of two classical components to one is also a problem, but it is easier spottable and less likely to happen.

Problem 4: Recreating an already existing software architecture. Publishing architectures which already exist under a different name is just problem 3 but with a whole architecture. It is easier to spot these if you know the available architectures, but not for newcomers to ITS development.

Problem 5: Not fully implement an architecture. Maybe the source of problem 3 and 4 is that software architectures are not fully implemented, but changed just as wanted. The reason could be because the architecture is misunderstood or was not examined carefully enough. Partially following an architecture without explicitly explaining/naming/categorizing it, also creates hidden changes. The classical architecture is so abstract that leaving a component out and only implementing what seems right can easily not make the system an ITS anymore. You should trust an architecture and implement it fully or not claim you adopted the architecture. The second option is to adapt an existing architecture to a new architecture and explain changes to it.

Problem 6: Missing rationale for decisions. Missing explanations of why a decision was made is very confusing when deciding for one out of many architectures. You always have to make compromises in software engineering. It is important to clarify why you chose one specific direction. This is especially important for the cons of an architecture, which are often neglected in current literature.

Problem 7: Unabstracted components. Including implementation specific components in an architecture will make it more difficult to reuse in a more generalized architecture. Components which are just a technological artifact should not be included in a more generalized architecture. Reuse is one of the main goals of generalized software architectures and this can only be achieved with abstract components. An example is [3] with components like "Jess" or "Protégé-2000" and arrows like "Client GUI Download with Java WebStart", which in itself is good to publish, but it is harder to reuse the architecture in other projects.

Problem 8: Diverging tutoring model definition. The tutoring model is the component with the most unique names. This is also the component with the most substantively distinct definition of its function. It has so many meanings that the specific function can not be known from the name. Often it does not even include storable data, but is just a composition of the teaching behavior of the system. The different functions a tutoring model, which were defined in the architectures, are: a database of errors students make, a prefabricated database of answers the ITS can give, a prefabricated learning path as options for the student, a generation of a student-profile as an evaluated summary of their actions, a direct, live analysis of the student's answers, a live generation of answers of the ITS, an adaption of already existing purpose-built answers to the situation and dialog, when help information should be shown, when/how to adapt already existing help to only focus on relevant information, recommendations of teaching material matching the students abilities, an adaption of teaching material to the student in difficulty and content, a generation of learning material in the kind of fill-in math equation tasks, and a management and control of the teaching process.

Problem 9: Informal definitions. For a very complex system like an ITS it is crucial to have accurate explanations to understand it the way it is meant. Often there is only prose text without many technical terms and an image with colorful rectangles, ellipses and arrows without an exact definition of the elements used. In software engineering it is expected to use a formal diagram style and well defined technical terms. An example of an ITS software architecture visualization with an informal style is [4].

Problem 10: Connections between components are neglected. There are many architectures which seem to have put less thought into the connections, which can be a problem when you focus too much on the elements and their functionality. The first indication is that the connections often do not have directions or all of them are bidirectional. The second is that almost all essential components have connections to each other. A sophisticated architecture may be able to reduce the number of connections for some components and also reduce the amount of complexity, misuse and therefore errors or misinterpretation. An example of an ITS software architecture that every essential component can talk to each other is [5].

Problem 11: Process in the architecture. A software architecture is not there to include a process description. Trying to shape the architecture in a sequential form so that a process is in the structure of the software itself makes it easy to understand the rough order of things. However, only very simple processes can be integrated that way and changing the process will be very difficult. It will neglect the more needed connections between components which are outside of the process. An example is the withdrawn architecture LTSA [6].

Problem 12: No ITS software architecture. Definitions that are not software architectures are hardest to implement. They likely model a part of the inner workings of an ITS, but are so abstract or do not focus on the structure that they do not count as a software architecture anymore. An example is the triangle model of the “Situation Model”, “Interaction Model” and “Affordance Model” [7].

Problem 13: Lacking development. Some of the architectures do develop an already existing architecture with very small changes. It creates unnecessary redundancy and an example is the [8], which just shows the architecture, but makes a new image and definitions.

4 Improvements to the Classical ITS Software Architecture

I suggest the use of a general ITS software architecture to improve the formal definition with the UML component diagram and also a fifth component which is the most used component and is not already in the classical ITS software architecture.

Improvement 1: Clear definition of the classic components. The naming should be precise and also specific to a software architecture. Three components represent knowledge which are necessary for an ITS: domain knowledge to teach, student knowledge to estimate the student and pedagogical knowledge on how to teach. The stored data is for performance reasons mostly not what one would expect, transformations are needed to get a more natural presentation of the knowledge, and an interpretation of the needed data. The functionality of these knowledge components is to manage the data about the different areas (domain, student, pedagogical) with saving, editing, deleting, filtering, transformations and low complex operations. The second step is to add the word "component", since it should be a clear difference between the underlying data and their managing components. The interface component is there for managing the user's visible and hearable interface on which the user can operate.

Improvement 2: Addition of a process steering component. We checked which functionality we found often as a new component in the architectures and found a component whose task is to control the communication between the components and decide what the ITS should do next. It knows the big picture of what is going on and can integrate a process without affecting the architecture. 34.8% of the architectures had a process steering component of any variation. Such a component is a good addition to the classical architecture.

We used UML component diagram to create an improved version of the classical ITS architecture with the added process steering component in a formal way. These improvements are built upon an architecture from 2003 [9], at its center is a component (named "tutoring process model") which acts as mediator between the interface component and the knowledge components. Since the abilities of the knowledge components are now more limited, the process steering component is now more powerful in and through extension than a tutoring process model, see Fig. 8. This style splits complex functionality additions from the existing components and allows explicit extension points to add new features through new components. This is a more modular approach, which leaves flexibility to grow the ITS.

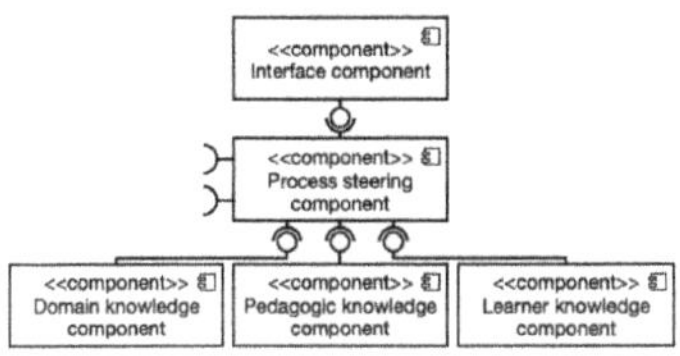

Fig. 8. Components of the classical ITS software architecture in a UML component model and the added process steering component

References

1. Nkambou, R., Mizoguchi, R., Bourdeau, J.: Advances in Intelligent Tutoring Systems. Springer, Heidelberg (2010). https://doi.org/10.1007/978-3-642-14363-2
2. Clancey, W.J.: Methodology for building an intelligent tutoring system. In: Methods and Tactics in Cognitive Science (1984)
3. Crowley, R., Medvedeva, O.: SlideTutor: a model-tracing intelligent tutoring system for teaching microscopic. Artif. Intell. Educ. Shaping Future Learn. Intell. Technol. **97**, 157 (2003)
4. Lee, Y., Cho, J., Choi, B.-U.: An intelligent tutoring system based on a multi-modal environment for the 300-certification program of english conversation. In: Ikeda, M., Ashley, K.D., Chan, T.-W. (eds.) ITS 2006. LNCS, vol. 4053, pp. 778–780. Springer, Heidelberg (2006). https://doi.org/10.1007/11774303_98

5. Kojima, K., Miwa, K.: Evaluation of a system that generates word problems through interactions with a user. In: Ikeda, M., Ashley, K.D., Chan, T.-W. (eds.) ITS 2006. LNCS, vol. 4053, pp. 124–133. Springer, Heidelberg (2006). https://doi.org/10.1007/11774303_13
6. IEEE: 1484.1-2003 - Learning Technology Systems Architecture (LTSA). Learning Technology Standards Committee of the IEEE Computer Society (2003)
7. Self, J.: The defining characteristics of intelligent tutoring systems research: ITSs care, precisely. Int. J. Artif. Intell. Educ. (IJAIED) **10**, 350–364 (1998)
8. Ahuja, N.J., Silla, R.: A critical review of development of intelligent tutoring systems: retrospect. Present Prospect. IJCSI **10**(4), 39 (2013). 1694-0814
9. Martens, A.: Centralize the tutoring process in intelligent tutoring systems. In: ICNEE (2003)

Assessing Students' Clinical Reasoning Using Gaze and EEG Features

Imène Jraidi[1(✉)], Asma Ben Khedher[1], Maher Chaouachi[2], and Claude Frasson[1]

[1] Department of Computer Science and Operations Research, University of Montreal, Montreal, QC H3T 1N8, Canada
{jraidiim, benkheda, frasson}@iro.umontreal.ca

[2] Department of Educational and Counselling Psychology, McGill University, Montreal, QC H3A 1Y2, Canada
maher.chaouachi@mcgill.ca

Abstract. The purpose of this work is to analyze the learners' visual and brain behaviors during clinical reasoning. An experimental study was conducted to record gaze and EEG data of 15 novice medical students as they interacted with a computer-based learning environment in order to treat medical cases. We describe our approach to track the learners' reasoning process using the visual scanpath followed during the clinical diagnosis and present our methodology to assess the learners' brain activity using the engagement and the workload cerebral indexes. We determine which visual and EEG features are related to the students' performance and analyze the relationship between the students' visual behavior and brain activity.

Keywords: Eye tracking · Scanpath · EEG · Engagement · Workload · Clinical reasoning · Learning performance

1 Introduction

The democratization of the use of affective computing techniques has enabled a multifaceted understanding of the learners' behavior [1, 2]. Particularly, these techniques allowed computer-based learning environments and intelligent tutoring systems to have access to a learner-centered data and to have a fine-grained analysis of the variables that impact learning performance [3–5].

Techniques such as electroencephalography (EEG) and eye tracking, which have been up to recent years mainly used in strict laboratory conditions, are being increasingly used in realistic learning settings [6–9]. Their capacity to offer real-time qualitatively rich information about the learners' state could be highly useful to optimize the learning strategies. This is particularly true in medical learning environments in which the learners' interactivity with the system is very limited [10, 11]. In fact, in such a clinical context, and throughout the diagnostic phases, the learners are generally engaged in various reading activities during which a complex clinical reasoning is performed with a very restrained action field. This consequently limits the learners' data acquired by the system and hinders its capacity to assess the learning process.

A. Coy et al. (Eds.): ITS 2019, LNCS 11528, pp. 47–56, 2019.
https://doi.org/10.1007/978-3-030-22244-4_7

Therefore, the integration of EEG sensors to track the learners' mental state, and eye tracking to track gaze patterns could be highly beneficial to assess learning, especially when the learner is just staring at a computer screen.

In this paper, we propose to use these techniques as a mean to analyze the learners' clinical reasoning. More precisely, we use EEG data to assess two mental states, namely engagement and workload, and eye tracking to monitor learner's visual focus and scanpath. We are interested in analyzing which features are related to the students' reasoning and performance.

The remainder of the paper is organized as follows. We present previous work on the use of EEG and eye tracking in the learning context. Next, we describe the developed tutoring system, experimental setup and methodology. Then, we discuss the obtained results, conclude and present directions for future work.

2 Previous Work

A wide range of sensing technologies is nowadays capable of providing highly accurate data on the users' state in human computer interactions and particularly in ITS, where the behavioral and affective component is being more and more important.

Specifically, eye tracking and EEG devices are being increasingly used within the learning context [12–16]. Slanzi and his colleagues predicted behavioral variables such as the learners' click intentions within a web search task using gaze fixation, pupil dilation and EEG [15]. Brouwer et al. used fixation duration, pupil size and EEG data to assess learners' attention in a visual search task [16]. El-Abbasy et al. developed an affective e-learning platform that changes the learning materials when the learner experiences a negative emotion. Emotions such as sadness and frustration were recognized using eye tracking and EEG [17]. In [18], Alhassan and his colleagues used data mining techniques to classify the learners in terms of two different learning styles, namely visual and verbal, by taking the participants' EEG and gaze data as input features. In the same context, in [19] the authors used machine learning algorithms to discriminate between high and low creativity students using skin conductance bracelet, an eye tracker and EEG sensors. Two classifiers namely, Nearest Neighbor and Naïve Bayes achieved up to 80% of true positive rate. Makransky et al. analyzed EEG measures, such as alpha and theta frequency band powers, and eye tracking metrics, such as the percentage of time spent on relevant regions of the interface, to assess learners' cognitive load during multimedia learning. Statistically significant relationships were found showing that eye tracking measures are indicators of extraneous cognitive load, and that EEG measures are indicators of intrinsic cognitive load [20].

In this paper, we propose to analyze the learners' visual behavior and brain activity within clinical reasoning using eye tracking and EEG. We use gaze and neural patterns to assess learners' visual focus and cognitive state, and determine which features are related to students' analytical process and performance.

3 Experimental Design and Methodology

An experimental study was conducted to record gaze and EEG data of novice medicine students as they interact with a computer-based learning environment (Amnesia). 15 participants (8 males) aged 20–27 years ($M = 21.8 \pm 2.73$) were recruited. Upon their arrival, they were briefed about the experimental procedure and asked to sign a consent form. They were then outfitted with an Emotiv EEG headset and placed in front of a Tobii TX-300 eye tracker (at 65 cm approximately). A sampling rate of 128 Hz was used to collect the EEG data from 14 different regions of the scalp (O1, O2, P7, P8, T7, T8, FC5, FC6, F3, F4, F7, F8, AF3 and AF4). A sampling rate of 300 Hz was used with the eye tracker, and a nine-point calibration grid was used to calibrate the participants' point of gaze.

3.1 Amnesia

Amnesia is a realistic environment developed to assess undergraduate medical students' analytical skills through clinical problem-solving [21]. The system features a virtual hospital where the learner acts as a doctor who is mistakenly diagnosed with amnesia. The learner needs to prove that he does not suffer from this disease by resolving six different medical cases. In each case, the learner is instructed to identify both the correct diagnosis and the appropriate treatment through a series of observations including the patients' demographic information, symptoms, antecedents and clinical data. The diseases he must find are respectively: flu, bacterial pneumonia, measles, Ebola, mumps and whooping cough. For each diagnosis and treatment, different response alternatives are given, and the student has up to three attempts to find out the correct answer.

3.2 Visual Behavior

Two different metrics were computed from the eye tracker, namely fixation duration and scanpath. Fixation duration is a temporal measure of the visual focus. It corresponds to the amount of time the fovea (center of gaze) is directed towards Areas of Interest (AOI) of the screen, i.e. the task-relevant elements in the system's interface. In total, six AOI were defined for each medical case (see Fig. 1 [22]):

- Information (I) includes the demographic information of the patient (e.g. name, age and weight).
- Antecedents (A) introduces the allergies and the prior diseases.
- Symptoms (S) highlights the symptoms of the disease.
- Analysis (N) includes data on the patient's temperature, heart rate and blood pressure.
- Diagnosis (D) shows different response suggestions for the disease to be identified.
- Treatment (T) presents different suggestions for the potential treatments of the disease.

The scanpath is a spatiotemporal measure representing the dynamic visual trajectory of the eye movements. It uses both fixations and saccades across the AOI and is recorded as a sequence (e.g. IIIAASSSSSNNNNSSNDDDSSDDTTDTT); each letter denotes a visited AOI with a fixation duration above a threshold of 250 ms. Redundant characters are then collapsed to retain a unique occurrence of each area (e.g. IASNSNDSDTDT). Smith-Waterman algorithm is used to evaluate this scanpath. The algorithm, which is notably used in bioinformatics for DNA sequence alignment, compares the learner's visual scanpath with a target reference sequence. This sequence, denoted as ISANDT, represents the hypothetico-deductive analytical process a novice clinician should follow in clinical reasoning [23]. In this process, the clinician starts by collecting the patient's information, and then formulates an initial hypothesis according to the symptoms. Next, additional clinical data such as analysis or medical antecedents are collected to validate or discard this hypothesis until reaching a final diagnosis, and a treatment is prescribed.

The algorithm aligns both sequences (the learner's scanpath and the reference sequence) by optimizing a *similarity score* derived from the following features: the number of *matches*, number of *mismatches* and number of *gaps*. Matches are convergent elements (i.e. identical letters) in both sequences. Mismatches are the divergent elements (requiring mutations, i.e. substituting one letter for another). Gaps are the missing elements (implying an insertion or a deletion in one of the two sequences); we refer to [14] for more details. The higher the aligned similarity score, the more the two sequences are similar; which means the closer the learner's visual sequence is to the optimal reference sequence.

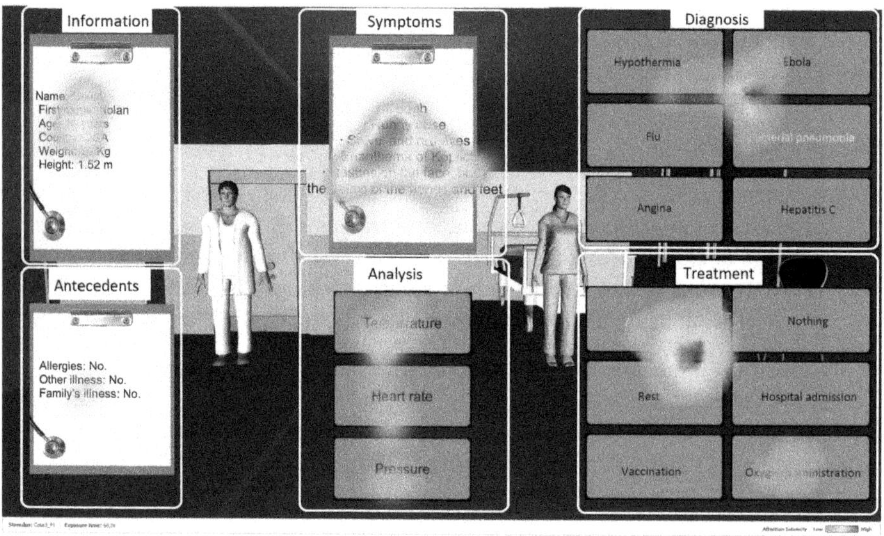

Fig. 1. Screenshot of amnesia with the AOI. The red color represents the most fixated areas, followed by yellow and then green with lower intensity. (Color figure online)

3.3 Brain Activity

Two brain indexes were computed from the EEG recordings, namely engagement and workload as depicted in Fig. 2 [24]. The engagement index is a neural indicator of the level of alertness and attention allocated during a task, and the workload index (also known as cognitive load index) measures the amount of information processing demands and mental effort induced during a task [6].

The engagement index is computed using three frequency bands, namely θ (4–8 Hz), α (8–13 Hz) and β (13–22 Hz) as follows: $\beta/(\theta + \alpha)$ [25]. The extraction of these bands is performed by multiplying one second of the EEG signal by a Hamming window in order to reduce the spectral leakage, and applying a Fast Fourier Transform (FFT). As the EEG headset measures 14 different regions at the same time, the values of θ, α and β are summed over all these regions. An engagement index is computed each second. Then, a 40-second moving average mobile window is used in order to smooth the index and reduce its fluctuation.

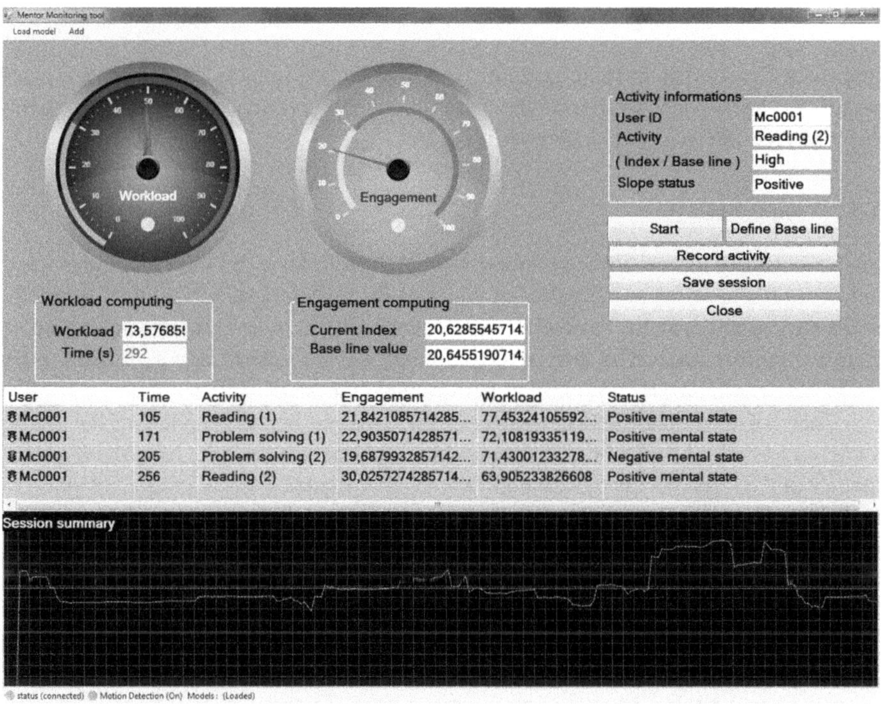

Fig. 2. Screenshot from the acquisition module used for the EEG data

Unlike the engagement index, which is directly extracted from the EEG raw data, the workload index is computed using a pre-trained and validated predictive model [26]. EEG signal is cut into 1-second segments and multiplied by a Hamming window. A FFT is then applied to transform each segment into a spectral frequency and generate a set of 40 bins of 1 Hz ranging from 4 to 43 Hz. The dimensionality of the data is reduced using a Principal Component Analysis and then mean normalized. Next, a Gaussian Process Regression algorithm with an exponential squared kernel and a Gaussian noise is run in order to compute the EEG workload index from the normalized data (we refer to [26] for more details about this procedure).

To assess the learner's mental state, a slope of each index is computed using the least squared error function of the indexes' values throughout each medical case. For the engagement index, if the slope value is positive, then the learner is considered as mentally *engaged*. Otherwise, the learner is considered as mentally *disengaged*. For the workload index, if the slope value is between −0.03 and +0.03, then the workload is considered as *optimal*. Otherwise, if the slope value is above 0.03, the learner is considered as *overloaded*, and if the slope is below −0.03 the learner is considered as *underloaded* [26]. Moreover, if the learner is mentally engaged and has an optimal workload, then the learner's mental state is considered as *positive*. Otherwise, the mental state is *negative*.

4 Results and Discussion

Results are presented in three sections: the first section analyzes the learners' visual behavior, the second section deals with the brain activity, and the third section describes the interplay between both measures.

4.1 Visual Behavior

First, we wanted to analyze the relationship between fixation duration and the learners' outcomes (correctness and number of attempts in each medical case). That is whether focusing (visually) at the task-relevant elements of the screen has an impact on the learners' performance. The answer is no: there was no statistically significant correlation between fixation duration and performance measures (p = n.s.).

Then, we were interested in analyzing the dynamic visual behavior of the learner, namely assessing the relationship between the performance measures and the alignment metrics of the learners' visual scanpath: numbers of matches, mismatches and gaps, and the similarity score between the used scanpath and the optimal reference sequence. First, with regards to the success/failure (i.e. correct diagnosis and treatment), we found that the number of mismatches was significantly lower ($F(1, 75) = 13.585$, $p < 0.001$) if the case is successfully solved ($M = 0.00$, $SD = 0.00$) compared to failed cases ($M = 0.21$, $SD = 0.41$). Also, the similarity score was significantly higher ($F(1, 75) = 5.879$, $p < 0.05$) for the succeeded cases ($M = 5.79$, $SD = 2.42$) compared to the unsolved ones ($M = 3.58$, $SD = 5.61$). This implies that for the correct answers, the alignment scores between the students' scanpath and the optimal reference sequence were significantly higher and the number of mismatches smaller.

In addition, statistically significant results were found for the number of attempts per case. A significant positive correlation was found between the number of mismatches and the number of attempts ($r = 0.36$, $p < 0.001$). The more mismatches or deviations between the learner's visual sequence and the reference sequence, the more they were attempts per case; i.e. the more the learners had trouble finding the right answer. A negative correlation with the score alignment was also found ($r = -0.26$, $p < 0.05$): the higher the alignment score, the lower was the number of attempts. This means that the more the learner's visual sequence was close to the optimal reasoning sequence, the lower was the number of attempts to answer. In other words, the more the learners' reasoning was 'optimal', the faster they could find the right answer.

Hence, the analysis of the gaze behavior showed that the fixation duration had no impact on the learning performance. A student can watch/fix the screen without being engaged in the task. This will be indeed confirmed through the analysis of the learners' brain data (EEG engagement index). The analysis of the dynamic visual behavior allowed us to track and evaluate the visual scanpath followed by the learner during his reasoning. The more the visual sequence was close to the hypothetico-deductive analytical process (the reference sequence), the better were the performance. This validates our approach to monitor the learners' analytical process through the analysis of the followed visual scanpath over the AOI of the game and its evaluation according to the optimal reference path using a sequence alignment algorithm.

4.2 Brain Activity

Similarly, we analyzed the relationship between the EEG variables (mental state, workload and engagement) and the performance measures (correctness and number of attempts in each medical case). We found a statistically significant association between correctness and the workload variable (underload, optimal load and overload), $\chi(2) = 7.445$, $p < 0.05$. In particular, more than half of the correct answers (54.10% of the cases) were associated to the state of optimal load (against 26.20% to overload and 19.70% to underload).

For the number of attempts, we found a statistically reliable relationship with the mental state variable (positive/negative): $F(1, 74) = 13.725$, $p < 0.001$. The number of attempts was significantly higher when the mental state was negative ($M = 3.84$, $SD = 0.96$), against ($M = 2.96$, $SD = 1.02$) if the state was positive. We recall that a negative mental state is associated with one of these states: disengagement, or-and underload/overload. Two other ANOVAs were conducted to study each index separately. There was no significant relationship between workload and the number of attempts. However, a significant relationship between the engagement variable (engaged/disengaged) and the number of attempts was found: $F(1, 74) = 17.27$, $p < 0.001$. This number was significantly higher if the learners were disengaged ($M = 3.95$, $SD = 0.89$) against ($M = 3.03$, $SD = 1.04$) if the learners were mentally engaged. That is as the engagement level dropped, the learners had more difficulty in solving the medical cases. Or conversely when the engagement index was high, the number of attempts to resolve the cases decreased. In other words, the more attentive they were, the easier they found the right answers.

To summarize, the two cerebral metrics seem to have an impact on the performance measures. The cognitive load had an impact on the failure/success of the medical cases, and the engagement index had an impact on the number of attempts per case. On one hand, we have a significant association between the optimal load interval of the workload index and the succeeded medical cases. On the other hand, we have a lower number of attempts for the positive values of the engagement index.

4.3 Relationship Between Gaze and EEG Data

We started by analyzing the relationship between the learners' mental state and the fixation duration in the medical cases. A first ANOVA showed that the fixation duration was on average significantly higher ($F(1, 74) = 5.99$, $p < 0.05$) when the mental state was negative ($M = 326.24$, $SD = 44.18$) vs. ($M = 302.82$, $SD = 38.62$) in case of positive mental state.

Two more ANOVAs were performed to analyze the engagement and the workload indices respectively. The first analysis showed that there was no statistically significant relationship between the engagement level and the fixation duration. On the other side, the analysis of the mental load index showed that there was a significant relation between fixation duration and (underload, optimal load and overload): $F(2, 73) = 4.275$, $p < 0.05$.

This shows that the fixation duration is not a good indicator of mental engagement. One can fix or look at an area of interest on the screen, without really lending focus or attention. This rather suggests that a high fixation time is a sign of overload and mental fatigue. Moreover, we found a significant negative correlation between the engagement and the workload indexes' values ($r = -0.295$, $p < 0.05$), which means that the higher the level of attention, the more the induced level of mental effort decreased. Inversely also, if the mental load increased, i.e. in case of mental fatigue, the level of attention decreased. This was observed not only during the resolution of the medical cases, but throughout all the interaction with the game ($r = -0.204$, $p < 0.05$).

5 Conclusion

This paper presented an experimental study that analyzes eye tracking and EEG features of novice medical students while they interact with a hospital simulation game. The goal was to assess their visual and cerebral behaviors during clinical reasoning using different medical cases and sensors to record their brain and gaze data.

The analysis of these data first led us to the conclusion that the fixation duration on the screen or on relevant elements of the environment is not a good indicator of performance (outcomes) or attention (EEG engagement), but rather a sign of mental fatigue (overload). However, the analysis of the dynamic visual behavior using the learners' scanpath across the different AOI enabled us to evaluate the analytical process followed by the learners during diagnostic reasoning. Gaze data were then correlated with performance on one hand, and brain measurements (mental state and engagement/workload indexes) on the other hand. Furthermore, the analysis of the EEG data showed that there were also statistically significant correlations between the different brain measurements and learners' performance.

Therefore, this paper allowed us to confirm that both sensors (eye tracking and EEG) are highly valuable sources of information for the monitoring of the students' external gaze behavior in terms of visual scanpath and also their internal physiological state in terms of cerebral activity during problem solving. Our future work is directed towards using both sensors to actively adapt the game's tutoring interventions according to the learners' visual behavior and brain activity.

Acknowledgments. This work was supported by NSERC (National Science and Engineering Research Council) and SSHRC (Social and Human Research Council) through the LEADS project. We also thank Issam Tanoubi from the University of Montreal for his collaboration in the design of the Amnesia environment.

References

1. Jraidi, I., Frasson, C.: Student's uncertainty modeling through a multimodal sensor-based approach. J. Educ. Technol. Soc. **16**, 219–230 (2013)
2. Hou, H.-T.: Integrating cluster and sequential analysis to explore learners' flow and behavioral patterns in a simulation game with situated-learning context for science courses: a video-based process exploration. Comput. Hum. Behav. **48**, 424–435 (2015)
3. Ben Khedher, A., Jraidi, I., Frasson, C.: Static and dynamic eye movement metrics for students' performance assessment. Smart Learning Environments **5**(1), https://doi.org/10.1186/s40561-018-0065-y (2018)
4. D'Mello, S.K., et al.: AutoTutor detects and responds to learners affective and cognitive states. In: Presented at the Workshop on Emotional and Cognitive Issues at the International Conference on Intelligent Tutoring Systems (2008)
5. Pardo, A., Han, F., Ellis, R.A.: Combining university student self-regulated learning indicators and engagement with online learning events to predict academic performance. IEEE Transact. Learn. Technol. **10**, 82–92 (2017)
6. Berka, C., et al.: EEG correlates of task engagement and mental workload in vigilance, learning, and memory tasks. Aviat. Space Environ. Med. **78**, B231–B244 (2007)
7. Ben Khedher, A., Jraidi, I., Frasson, C.: Tracking students' mental engagement using EEG signals during an interaction with a virtual learning environment. J. Intell. Learn. Syst. Appl. **11**, 1–14 (2019)
8. Maynard, O.M., Munafò, M.R., Leonards, U.: Visual attention to health warnings on plain tobacco packaging in adolescent smokers and non-smokers. Addiction **108**, 413–419 (2013)
9. Ben Khedher, A., Jraidi, I., Frasson, C.: What can eye movement patterns reveal about learners' performance? In: 14th International Conference on Intelligent Tutoring Systems (ITS 2018). LNCS, vol. 10858, pp. 415–417. Springer (2018)
10. Poitras, E.G., Doleck, T., Lajoie, S.P.: Towards detection of learner misconceptions in a medical learning environment: a subgroup discovery approach. Educ. Tech. Res. Dev. **66**, 129–145 (2018)
11. Lajoie, S.P., Naismith, L., Poitras, E., Hong, Y.-J., Cruz-Panesso, I., Ranellucci, J., Mamane, S., Wiseman, J.: Technology-rich tools to support self-regulated learning and performance in medicine. In: Azevedo, R., Aleven, V. (eds.) International Handbook of Metacognition and Learning Technologies. SIHE, vol. 28, pp. 229–242. Springer, New York (2013). https://doi.org/10.1007/978-1-4419-5546-3_16
12. D'Mello, S., Olney, A., Williams, C., Hays, P.: Gaze tutor: a gaze-reactive intelligent tutoring system. Int. J. Hum Comput Stud. **70**, 377–398 (2012)
13. Lallé, S., Conati, C., Carenini, G.: Predicting confusion in information visualization from eye tracking and interaction data. In: Proceedings of the Twenty-Fifth International Joint Conference on Artificial Intelligence, pp. 2529–2535. AAAI Press (2016)
14. Ben Khedher, A., Jraidi, I., Frasson, C.: Local sequence alignment for scan path similarity assessment. Int. J. Inf. Educ. Technol. **8**(7), 482–490 (2018). https://doi.org/10.18178/ijiet.2018.8.7.1086

15. Slanzi, G., Balazs, J., Velasquez, J.: Combining eye tracking, pupil dilation and EEG analysis for predicting web users click intention. Inf. Fusion **35**, 51–57 (2017). https://doi.org/10.1016/j.inffus.2016.09.003
16. Brouwer, A.-M., Hogervorst, M.A., Oudejans, B., Ries, A.J., Touryan, J.: EEG and eye tracking signatures of target encoding during structured visual search. Front. Hum. Neurosci. **11**, 264 (2017). https://doi.org/10.3389/fnhum.2017.00264
17. El-Abbasy, K., Angelopoulou, A., Towell, T.: Measuring the Engagement of the Learner in a Controlled Environment using Three Different Biosensors. Presented at the 10th International Conference on Computer Supported Education February 8 (2019)
18. Alhasan, K., Chen, L., Chen, F.: An experimental study of learning behaviour in an elearning environment. In: The IEEE 20th International Conference on High Performance Computing and Communications, pp. 1398–1403 (2018)
19. Muldner, K., Burleson, W.: Utilizing sensor data to model students' creativity in a digital environment. Comput. Hum. Behav. **42**, 127–137 (2015)
20. Makransky, G., Terkildsen, T.S., Mayer, R.E.: Role of subjective and objective measures of cognitive processing during learning in explaining the spatial contiguity effect. Learn. Instr. **61**, 23–34 (2019)
21. Ben Khedher, A., Jraidi, I., Frasson, C.: Tracking students' analytical reasoning using visual scan paths. In: 17th IEEE International Conference on Advanced Learning Technologies (ICALT), pp. 53–54. IEEE (2017)
22. Ben Khedher, A., Jraidi, I., Frasson, C.: Exploring students' eye movements to assess learning performance in a serious game. In: EdMedia + Innovate Learning: Association for the Advancement of Computing in Education, pp. 394–401. AACE (2018)
23. Swanson, H.L., O'Connor, J.E., Cooney, J.B.: An information processing analysis of expert and novice teachers' problem solving. Am. Educ. Res. J. **27**, 533–556 (1990)
24. Chaouachi, M.: Modélisation de l'engagement et de la charge mentale de travail dans les Systèmes Tutoriels Intelligents. Ph.D. thesis, Université de Montréal (2015). https://papyrus.bib.umontreal.ca/xmlui/handle/1866/11958
25. Pope, A.T., Bogart, E.H., Bartolome, D.S.: Biocybernetic system evaluates indices of operator engagement in automated task. Biol. Psychol. **40**, 187–195 (1995)
26. Chaouachi, M., Jraidi, I., Frasson, C.: Modeling mental workload using EEG features for intelligent systems. In: User Modeling, Adaption and Personalization, pp. 50–61 (2011)

Computer-Aided Intervention for Reading Comprehension Disabilities

Chia-Ling Tsai[1(✉)], Yong-Guei Lin[2,3], Wen-Yang Lin[2], and Marlene Zakierski[4]

[1] Iona College, New Rochelle, NY 10801, USA
ctsai@iona.edu
[2] Chung Cheng University, Minxiong Township, Chiayi, Taiwan
[3] National Yunlin University of Science and Technology, Yunlin, Taiwan
[4] Sage Colleges, Albany, NY 12209, USA

Abstract. Our research work focuses on grouping of students based on error patterns in assessment outcomes for effective teaching of reading comprehension in early elementary education. The work can facilitate placement of students with similar reading disabilities in the same intervention group to optimize corrective actions. We collected ELA (English Language Arts) assessment data from two different schools in NY, USA, involving 365 students in total. To protect individual privacy of the participants, no background information that can possibly lead to their identification is collected for the study. To analyze underlying factors affecting reading comprehension without students' background information and to be able to evaluate the work, we transformed the problem to a K-nearest neighbor matching problem—an assessment should be matched to other assessments performed by the same student in the feature space. The framework allows exploration of various levels of reading skills as the features and a variety of matching mechanisms. In this paper, we present studies on low-level features using the computer-generated measures adopted by literacy experts for gauging the grade-level readability of a piece of writing, and high-level features using human-identified reading comprehension skills required for answering the assessment questions. For both studies, the matching criterion is the distance between two feature vectors. Overall, the low-level feature set performs better than the high-level set, and the difference is most noticeable for K between 15 and 30.

Keywords: Reading comprehension · Intervention program · Feature matching

1 Introduction

The Common Core Learning Standards (CCLS), adopted by forty-four U.S.A. states and the District of Columbia define the knowledge and skills that a student should demonstrate by the end of each grade [1]. One important skill emphasized by CCLS is the reading ability, which is the precursor for learning in all content areas, including

This work was supported by NSF, under grant NSF-CHS-1543639.

A. Coy et al. (Eds.): ITS 2019, LNCS 11528, pp. 57–62, 2019.
https://doi.org/10.1007/978-3-030-22244-4_8

science, technology, engineering, and mathematics (STEM). Being able to read independently and analytically enables critical thinking and is essential to students' future success in the globally competitive workforce in the 21st century.

In New York State, students in grades 3–8 take the State English Language Arts (ELA) test, which measures the CCLS. An ELA test consists of multiple choice questions and open-ended questions based on short passages in the test. To do well, students should be able to read the text closely for textual evidence and to make logical inferences from it. To report the results, the number of correct responses a student answers on a test is converted into one single scale score to be mapped to the corresponding performance level. However, a single performance score as the outcome measure is often insufficient in identifying underlying learning problems for intervention, especially for reading comprehension [2].

The goal of this study is to group students based on error patterns in assessment outcomes at an individual level to facilitate effective teaching for reading comprehension in early elementary education. Similar research can be found in the area of personalized speech disorder therapy [3, 4]; the system makes predictions on the outcome of a therapy based on the big volume of data collected from patients who have completed the therapy program. To the best of our knowledge, we have not yet come across similar research work for personalized reading comprehension intervention.

2 Methodology

The dataset consists of three parts. The first part is the set of 6 fourth grade New York State mock ELA examinations from year 2005 to 2010. Only multiple-choice questions are considered and each question is labeled with one of 9 performance indicators defined by New York State Education Department (NYSED), covering 3 main high-level reading skills—making summary, making inference, and locating information. Such labels are provided by a literacy expert. The second part is the set of student assessments, involving a total of 365 fourth grade students from 17 reading intervention classes from 2 different schools in NY. Every participant was assigned an identification number and should participate in three mock examinations. No other background information, such as prior academic performance or demographic data, was collected to protect the privacy of the participants. The third part is a dictionary compiled from a collection of books recommended by NYSED to represent the lexical knowledge base for children up to fourth grade. Information encoded in the dictionary include word frequency, phonetic spelling (Arpabet [5]), and phonotactic information [6].

2.1 Problem Formulation

The collected data on student assessments poses a fundamental challenge—no groundtruth information on students' prior reading performance available for validation. More specifically, students' reading disabilities (defined as lack of certain fundamental cognitive skills) are unknown. In addition to the challenge, the data also suffers from having low signal to noise ratio, since students are known to make random choices in an examination when feeling lost, and having incomplete records, since not all students participated in all three mock examinations.

To overcome the fundamental challenge of validating correct grouping of students with similar reading disabilities, we make an assumption about a student having similar performance in multiple assessments, so one assessment should be grouped with other assessments done by the same student, since the only piece of groundtruth information available is the temporal relationship among assessments done by the same student. We transform the problem to a K-Nearest-Neighbor (KNN) matching problem: given the right feature selection and a robust matching mechanism, if there are multiple assessments from the same student taken few months apart, each assessment should be a considerably close match to other assessments of the same student in the feature space. When evaluating the work, if $K = 1$ and the closest match is indeed an assessment done by the same student, the match is considered a success, else it is a failure. The framework allows exploration of features in terms of reading skills, and various matching mechanisms. In this paper, we report our preliminary results from two different feature sets: (a) high-level and manually labeled, and (b) low-level and computer-generated. The matching criteria is the Mahalanobis distance in the feature space.

2.2 Feature Selection

To generate the high-level feature set, a literacy expert was asked to label each examination question using one of the 9 identifiable reading skills: main idea, author's purpose, conclusion, inference, context clue, prediction, sequence, interpretation and locating information. A question is labeled with a specific skill if the skill is determined by a literacy expert to be the major skill required to correctly answer the question. Due to the small number of questions available in each examination (no more than 28), we might not have enough questions to cover all 9 reading skills. To avoid such a problem, we reduce the number of features to 3: the first 3 reading skills are consolidated into one category for making summary, the next 3 for making inference and the last 3 for locating information. Each question is given a 3-tuple binary feature vector, with 1 in the position of the category that the question belongs to, and 0 in all other positions. For example, if the question is labeled "conclusion", it falls under "making summary" category and the feature vector is $[1, 0, 0]$. To generate a feature vector for a student assessment, the feature vectors of all correctly answered questions are added, and each ith element is normalized by the total number of questions in the ith category. Reducing the number of reading skills from 9 to 3, we also increase the distinguishing power in each feature dimension for student assessment.

To determine the low-level feature set of a student assessment, we first compute such features for each question. To do so, we follow the same guide that the New York Education Department adopted for computing the quantitative measures of text complexity of a piece of article for being grade-appropriate in ELA. Such measures include average word length (number of syllables), average sentence length (number of words), word frequency, and text cohesion. Many studies have supported the predicting powers of those measures [7–10]. To estimate text cohesion, we first convert a sentence to a speech graph [10], with each word being represented with its part-of-speech (PoS) tag as a node, and the outgoing edge connecting to the PoS tag of the next word. A graph contains at most one node per PoS tag, so different words of the same tag share the same node, resulting in loops in the graph. The number of loops in a speech graph measures the text cohesion of a sentence. We extract the tags using Stanford Log-linear Part-Of-Speech Tagger system [11]. To generate a feature vector for a student assessment, we add all the feature vectors of correctly answered questions and subtract all the feature vectors of incorrectly answered questions. Because not all student assessments possess the same number of questions (due to absence or other reasons), the outcome vector is normalized by the number of applicable questions in the assessment.

2.3 Matching

The dissimilarity function is the distance between two vectors in the feature space. To take into consideration the variance in each feature dimension, the dissimilarity function is defined as the Mahalanobis distance:

$$d = (x, y) = \sqrt{(x-y)^T C^{-1} (x-y)},$$

where x and y are two feature vectors. C is the covariance matrix, computed as:

$$C = \frac{1}{n-1} \sum_{i=1}^{n} (v_i - \omega)^T (v_i - \omega),$$

where ω is the average feature vector. Depending on the type of feature set, $\{v_i\}$ is sampled differently. For the high-level feature set, $\{v_i\}$ is the set of feature vectors of student assessments of the other two examinations. For the low-level feature set, the population for deriving the covariance matrix consists of all the multiple-choice questions and the original articles broken up into short passages of about 100 words. Each question/passage is assigned a feature vector v_i. For both types of feature sets, ω is computed as the mean of $\{v_i\}$.

3 Results and Evaluation

The total number of students participated in the study is 365. However, only 281 attended all 3 mock examinations. Given a student assessment, it is matched against assessments of the other two examinations done by the same class. To evaluate the work, a match is considered successful if at least one of the two other assessments performed by the same student is matched within K nearest neighbors. Since the maximum class size is 22, yielding 44 neighbors to be matched from the other two examinations, and we wish to consolidate the results from the 17 classes, we randomly add assessments from other classes to reach 44 for classes having fewer than 22 students. There is a total of 843 cases considered, three from each student. Table 1 shows the success rates of KNN matching using both types of feature sets, with K up to 8, and Fig. 1 shows the complete results for K going from 1 to 44.

Table 1. Success rates of KNN matching with $K <= 8$ using different feature sets

K	1	2	3	4	5	6	7	8
High-level feature set (%)	8.82	17.38	23.80	30.39	36.19	41.89	46.26	51.16
Low-level feature set (%)	10.43	18.81	24.78	30.84	36.19	42.60	47.77	52.58
Difference (%)	1.61	1.43	0.98	0.45	−0.09	0.71	1.51	1.42

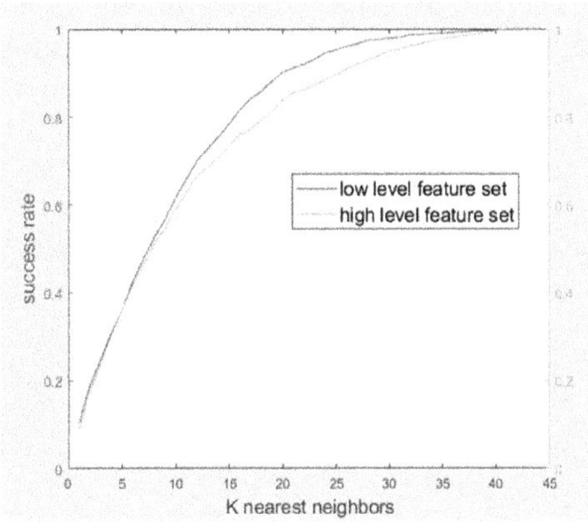

Fig. 1. Complete results of KNN matching using high-level and low-level feature sets, respectively.

Overall, the low-level feature set performs better than the high-level feature set, and the difference is most noticeable for K between 15 and 30. Possible explanations for this outcome is high subjectivity in the labels provided by a human expert, and inappropriate representation of the high-level reading skill feature space since a question often demands a combination of multiple reading skills.

4 Discussion and Future Work

The work is still in its early infancy, as the best preliminary results using the popular low-level features adopted by the education community still show plenty of room for improvement. However, the low accuracy might be partially attributed to inaccurate assumption about students maintaining similar performance in multiple assessments, and noise in the data set, as mentioned earlier. For the former, it violates the purpose of intervention, since students are expected to improve in the assessment score if the intervention is effective. On average, the difference between the numbers of correctly answered questions from two assessments by the same student is about 3.7 (with a total of 28 questions). 11.7% of the participants show an improvement of 8 more correctly-answered questions in later assessments. Regarding the noise in the data, 6.8% of the participants show a deterioration of 8 more incorrectly-answered questions in later assessments, which can hardly be explained as sensible behaviors. In addition, it is also impossible to verify whether students actually applied the same identification numbers as instructed in all three mock examinations.

Our preliminary work confirmed the superiority of computer-generated features over manually-labeled features in locating similar error patterns in assessment outcomes. The next step is to continue the search for more effective features, and also matching mechanisms. For the latter, possible directions include question-based matching between every two assessments with the constraints of a bipartite graph.

We anticipate that any success in this work can lead to more effective grouping of students with similar reading disabilities in the same intervention group to optimize corrective actions and use of school teaching resources. The long-term vision is the development of an intelligent Response-to-Intervention system that delivers an instruction plan as a function of the outcome of an individual assessment, by matching to a big volume of student assessment data, to facilitate efficient reading intervention.

References

1. Common Core State Standards Initiative. http://www.corestandards.org/. Accessed 18 Mar 2019
2. Wren, S.: The Cognitive Foundations of Learning to Read: A Framework. Southwest Educational Development Laboratory, Austin (2001)
3. Danubianu, M., Socaciu, T.: Does data mining techniques optimize the personalized therapy of speech disorders? J. Appl. Comput. Sci. Math. **5**(3), 15–18 (2009)
4. Danubianu, M., Pentiuc, S.G., Tobolcea, I., Socaciu, T.: Model of a data mining system for personalized therapy of speech disorders. J. Appl. Comput. Sci. Math. **6**(3), 28–32 (2009)
5. Klatt, D.H.: Review of the ARPA speech understanding project. J. Acoust. Soc. Am. **62**, 1345 (1977)
6. Storkel, H.L., Hoover, J.R.: An on-line calculator to compute phonotactic probability and neighborhood density based on child corpora of spoken American English. Behav. Res. Methods **42**(2), 497–506 (2010)
7. Fry, E.: A readability formula that saves Time. J. Read. **11**, 513–516 (1968)
8. Graves, M.F., Boettcher, J.A., Peacock, J.L.: Word frequency as a predictor of students' reading vocabularies. J. Read. Behav. **12**(2), 117–127 (1980)
9. Ryder, R.J., Slate, W.H.: The relationship between word frequency and word knowledge. J. Educ. Res. **81**(5), 312–317 (1988)
10. Mota, N.B., et al.: Speech graphs provide a quantitative measure of thought disorder in psychosis. PLoS ONE **7**(4), e34928 (2012)
11. Stanford Log-linear Part-of-Speech Tagger. https://nlp.stanford.edu/software/tagger.shtml. Accessed 24 Mar 2019

Conceptualization of IMS that Estimates Learners' Mental States from Learners' Physiological Information Using Deep Neural Network Algorithm

Tatsunori Matsui[1(✉)], Yoshimasa Tawatsuji[1], Siyuan Fang[2], and Tatsuro Uno[3]

[1] Faculty of Human Sciences, Waseda University, 2-579-15 Mikajima, Tokorozawa, Saitama 359-1192, Japan
matsui-t@waseda.jp, wats-kkoreverfay@akane.waseda.jp
[2] Global Education Center, Waseda University, 1-6-1 Nishiwaseda, Shinjuku, Tokyo 169-8050, Japan
[3] School of Human Sciences, Waseda University, 2-579-15 Mikajima, Tokorozawa, Saitama 359-1192, Japan
katappo@fuji.waseda.jp

Abstract. To improve the efficiency of teaching and learning, it is substantially important to know learners' mental states during their learning processes. In this study, we tried to extract the relationships between the learner's mental states and the learner's physiological information complemented by the teacher's speech acts using machine learning. The results of the system simulation showed that the system could estimate the learner's mental states in high accuracy. Based on the construction of the system, we further discussed the concept of IMS and the necessary future work for IMS development.

Keywords: Intelligent mentoring system · Physiological information · Deep neural network

1 Introduction

1.1 Research Background and Objective

To improve the efficiency of teaching and learning, it is substantially important to know learners' mental states during their learning processes. Researches in educational technology has provided us with much knowledge on the relationships among learners' physiological information, such as eye motion and amount of sweat, and learners' learning behaviors and mental states. As today's computers are becoming more and more sophisticated in function and lower in price, they can be utilized to perform a great deal of real-time processing of human physiological data. Therefore, many researchers are developing learning-assisting systems that can automatically estimate learners' mental states. Furthermore, it is generally acknowledged that during the process of teaching and learning, the interaction between the teacher and the learners

A. Coy et al. (Eds.): ITS 2019, LNCS 11528, pp. 63–71, 2019.
https://doi.org/10.1007/978-3-030-22244-4_9

greatly influences the learners' mental states, and thereby is a determining factor of the learning effect. Hence, it is also vital to clarify the relationships between teachers' speech and behaviors, on one side, and learners' mental states and the factors that influence learners' mental states, on the other side. Such knowledge is expected to be integrated into the learning-assisting systems. The present authors especially focus on applying learning-assisting systems to e-learning learning environments, aiming at developing systems that can adapt to individual learners. As an important feature of such systems, in a teaching-and-learning scene, the human teacher (or an intelligent agent that plays the role of a teacher) and the learning-assisting system co-work in a complementary manner. Specifically, the learning-assisting system can provide the teacher with information that the teacher can hardly obtain, e.g., the learner's physiological information. The teacher can thus make teaching strategies by using observable information, e.g., the learner's behavioral information, complemented by the information provided by the system.

The relationships between learners' mental states and teachers' speech complemented with learners' physiological data were investigated by Takehana and Matsui [1]. Specifically, they tried to formalize the relationships concerning various learning-related information, such as physiological information, speech acts, and retrospective reports using the association rule mining technique. On the other hand, recent trends show the possibility to apply machine learning algorithms in learning assisting [2–4]. Thus, this study probes into the possibility of estimating learners' mental states from abovementioned multifaceted learning-related information using a deep neural network (DNN).

1.2 Structure of the Study

Section 2 introduces the Intelligent Mentoring Systems (IMSs) and the position of the present study. Section 3 describes the experiment through which we obtained various information about the learners, which was the object of the analyses in this study. In this experiment, we recorded a student's physiological information, a teacher's teaching behaviors (i.e., the teacher's speech), and the student's mental states. In Sect. 4, we constructed a multi-layer neural network which took the student's physiological information and the teacher's teaching behaviors as inputs, and the student's mental state as output. The results of the learning by the neural network were also described and discussed in this section. Section 5 summarizes the study and introduces our future work.

2 Intelligent Mentoring System

2.1 AutoTutor

D'Mello, Grasser, and Picard's [5] study is an important attempt to examine learners' mental states (*affective states*) in the field of intelligent tutoring systems (ITSs). They developed a system named AutoTutor that collected the learner's multifaceted behavior information (body pressure, eye-tracking, and facial features) and dialogue during a learning process and estimated the learner's mental states. Although the AutoTutor study is fairly important as it introduced the idea of estimating learners' mental states into the realm of ITSs, AutoTutor used only learners' behavioral information such as postures, eye-tracking information and facial features and did not make use of learners' physiological information which was considered to be a vital source of information for teachers' strategy making. The behavioral information is observable to the teachers, but the physiological information is unobservable. If teachers can use the behavioral information and the physiological information in the meantime, they will be able to estimate learners' mental states from more aspects. For this reason, it is of great importance to delve into the relationships between physiological information and mental states.

2.2 Outline of the Intelligent Mentoring System Proposed in This Study

This section introduces the IMSs. The present study is a fundamental research about the estimating mechanism through which an IMS can estimate learners' mental states.

Tatsunori Matsui's research group aims to develop models and fundamental techniques that are necessary to the realization of automatic mentoring systems that can estimate learners' states of knowledge comprehension and learners' mental states in e-learning scenes [3, 6]. Such learning-assisting systems are called Intelligent Mentoring Systems (Fig. 1). One feature of IMSs is that the diagnostic module of the learner model uses data on learners' mental states. Because learners' mental states change from time to time, it is necessary to monitor them and provide immediate feedbacks. In other words, IMSs fulfill their tasks of learning assisting by integrating the decisive models used in techniques of knowledge comprehension diagnosis and in methods for teaching assisting, which already exist in the studies of Intelligent Tutoring Systems (ITSs), and the decisive models used in real-time mental state estimation and the assisting techniques based on the estimation. This implies the necessity of two new technical foundations—a module that estimates learners' mental states and a module that decides the assisting methods based on the results of the diagnoses of learners' mental states.

In Matsui's research group's previous studies, the module that estimates learners' mental states were developed as a system that could automatically estimate learners' mental states in real-time. In order to apply the system in every-day PC-using environments, they avoided using apparatus that was difficult to install and operate. For example, although pupil size was one useful indicator of learners' mental states, it needed very complicated apparatus to measure it. They chose to employ low-level interaction (LLI) resource, which were behavioral features of learners.

Interactions could be sampled in various levels of granularity. Ryu and Monk [7] viewed interactions through GUIs as cyclic information exchanges between users and systems, and they termed the smallest units of such interactions as low-level interactions. Based on this definition, Matsui and his colleagues defined LLI resource as features of learners' behaviors such as changes in PC mouse movement speed, intervals between key pressings on keyboards, and changes in body posture which were sampled in the finest granularity. On the other hand, the output text strings and the time that learners took to fulfill their tasks were defined as high-level interaction (HLI) resource because their sampling granularity was relatively coarse compared to the LLI resource.

The HLI resource was high-level interaction information which was produced consciously by the learners, while the LLI resource was low-level interaction information which was produced unconsciously by the learners. Thus, estimating learners' mental states from learners' unconscious behaviors is one characteristic of IMSs.

Following this idea, Matsui and his colleagues made attempts to estimate learners' mental states in scenes of learning using LLI resource. For instance, as a fundamental study to realize the mental state estimation function in IMSs, they developed a system to estimate learners' degrees of confusion based on the movements of learners' PC mouses, the tilting angles of learners' heads, and the back-and-forth movements of learners' heads during their learning processes [3]. They also developed a system that could estimate learners' degrees of confidence from learners' eye movements when the learners were answering choice questions.

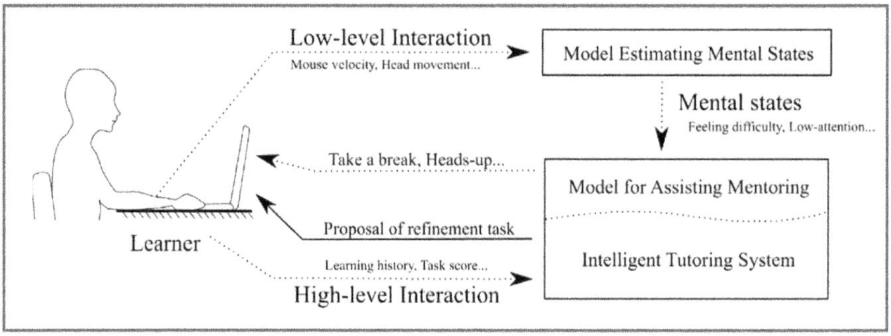

Fig. 1. Conceptual scheme of an intelligent mentoring system.

3 Collection of Multifaceted Learning-Related Information

In this study, we conducted an experiment to obtain multifaceted learning-related information using biometric devices. One teacher and one student in a private tutoring school participated in the experiment. Informed consent was obtained from the student and the student's guardians through the teacher. We measured the NIRS (Hitachi WOT-100), the pulse volume, the respiration intensity, and the skin conductance of the student, as well as the pulse volume, the respiration intensity, and the skin conductance of the teacher during the experiment. The pulse volume, the respiration intensity, and the skin conductance were measured using NeXus. The participants were asked to conduct a class in the same manner as in the private tutoring school. The beginning moment and end moment of the class were marked in the recordings of the devices.

The proceeding of the experiment was recorded by three video cameras located in three different positions in the experiment room. After the experiment finished, we divided the teacher's speech acts into nine categories– Explaining, Questioning, Comprehension Checking, Repeating, Praising, Alerting, Task Fulfillment Checking, Chatting, and Others, while watching the video records. This set of categories was previously used by Fujinoe [8], Kishi and Nojima [9] and Shimizu and Uchida [10], and revised by the present authors in this study. On another day, the student was asked to report how his mental states changed during the class while watching the video records. Furthermore, the teacher was asked to estimate how the student's mental states changed during the class. According to the Achievement Emotions Questionnaire (AEQ) [11], we divided the student's mental states into nine categories—Enjoy, Hope, Pride, Anger, Anxiety, Shame, Hopelessness, Boredom, and Relief. The student reported no mental state of Boredom, so we removed the category of Boredom and put the rest eight categories into the data analyses. As a result, we found that the consistency between the student's report and the teacher's estimation was 24.11% across the eight categories of mental states.

4 Analyses by Deep Neural Network

This section introduces the DNN system mentioned in Sect. 1 that maps the student's physiological information to the student's mental states. The system has five inputs, namely, the student's skin conductance, respiration intensity, pulse volume, and NIRS data, as well as the teacher's speech acts. The output of the system is the student's mental state, which has eight categories. We standardized the sampling rates, namely the granularities, of the inputs. However, to improve the real-world applicability of the system, we did not take data normalization and the delay of the changes of mental states into consideration, and did not preprocess the NIRS data using the global average reference method.

The neural network consisted of an input layer, four hidden-layers and an output layer. The number of the units on the hidden layers were determined in an exploratory manner. Specifically, we ran a simulation in which the numbers of units on the first, second and third hidden layers were set to be 69, 89 and 80, and the number of units on the fourth hidden layer changed gradually from one to 100. Figure 2 shows how the loss and accuracy changed as the simulation proceeded. The results of the simulation indicate that the optimal number of units on the fourth hidden layer is 69.

The DNN system was implemented using Tensorflow (ver 0.12.1) in Python 3.5. Each mapping was trained 70000 times. The activation function of the hidden layers was the tanh function, and the activation function of the output layer was the softmax function. The cost function was the cross-entropy error function, and the optimization method was the gradient descent. The learning rate was set to be 0.08. A cross-validation was run 10 times with 60% of the data being used as training data and the rest 40% being used as validation data. Figure 3 shows how the loss and accuracy changed during the training and the cross validation. Since the loss value converged greatly with this number of times of learning, it is reasonable to consider that the system had successfully acquired the mapping from the input information to the output information. The simulation results revealed that the system estimated the student's mental state in an accuracy of 76.17%, which was far higher than the accuracy of the human teacher's prediction, namely, 24.11%. This suggests that the DNN system is able to provide support in teaching and learning.

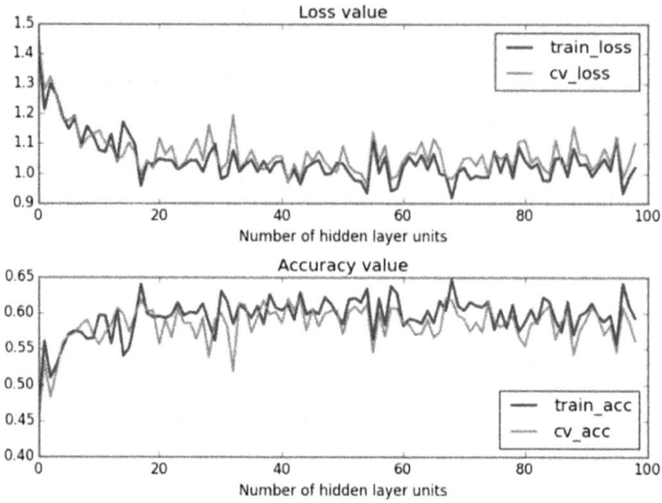

Fig. 2. Changes of loss and accuracy as the number of units on the fourth hidden layer changed from one to 100 while the numbers of units on the first, second and third hidden layers were set to be 69, 89 and 80.

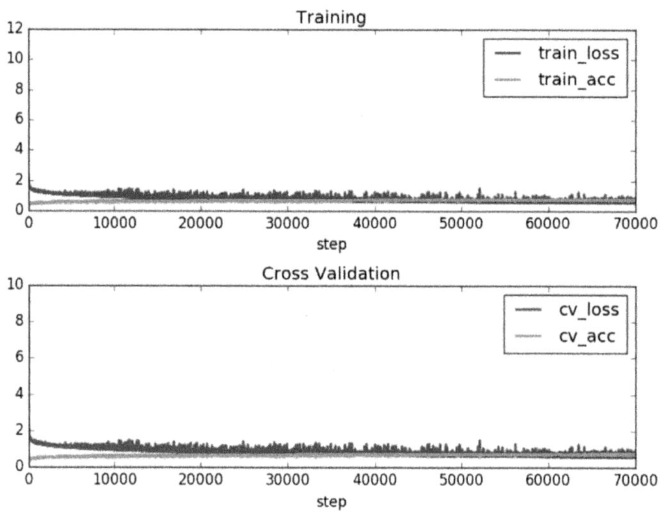

Fig. 3. Changes of loss and accuracy during the training and cross-validation of the deep learning system. The criterion of prediction accuracy is whether the predicted mental state and the actual mental state fall into the same category.

5 Towards the Realization of a Real-Time Mental State Estimation System

This section describes how the DNN system constructed in this study contributes to the construction of IMSs, as well as the issues to be addressed in the future. Figure 4 is a conceptual scheme of IMS that can estimate learners' mental states in real time using machine learning. This system takes the learner's physiological information and the teacher's speech acts as inputs, and feedback the learner's mental states to the teacher in real time. Because the physiological information that the system uses is LLI resource, which the teacher can hardly analyse, the information on the learner's mental states provided by the system can complement the teacher's comprehension of the learner's mental states. The present study shows that the DNN system can effectively estimate the learner's mental states from the learner's physiological information. The feedback of the estimation results to the teacher is effective, for example, when it is difficult for the teacher to read the mental states of the learner, and, therefore, the system is expected to provide the teacher with an opportunity to examine the teaching strategy diversely.

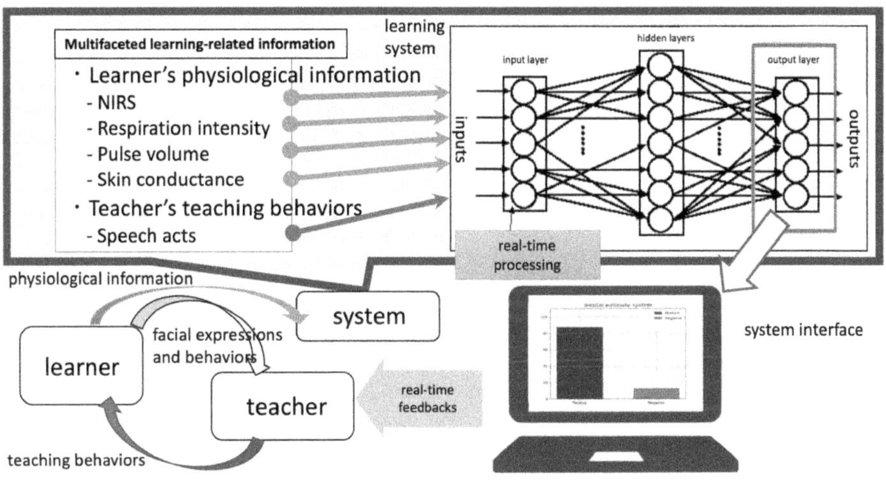

Fig. 4. Conceptual scheme of IMS that can estimate learners' mental states in real time using machine learning.

In order to achieve these objectives, it is of utmost importance to construct a mechanism that collects information in real time and feedbacks its estimation results. In addition, it is necessary to experimentally investigate how the teacher employs these real-time feedbacks in his/her teaching strategies. This work has been undertaken in Tawatsuji, Uno, Fang and Matsui's study [12]. Furthermore, it is necessary to find a way to generate readable descriptions on how the system estimates learners' mental states from learners' physiological information. In other words, it is necessary to interpret the causal relationships encoded by the system in the context of learning and education. From this point of view, the visualization and interpretation of the synaptic weights within the hidden layers are very important research subjects. We believe that the method for interpreting association rules used by Takehana and Matsui [1] could be one solution to the problem.

6 Summary

Our present research plan focuses on investigating learners' physiological information that is unobservable to human teachers in individual e-learning environments, aiming at constructing a system that can estimate learners' mental states. As the first step of the plan, this study tried to estimate a student's mental states from multifaceted learning-related information, namely the student's physiological information and the teacher's speech acts using machine learning. Specifically, we used DNN to extract the relationships between the student's mental states and the student's physiological information complemented with the teacher's speech acts. A simulation of the DNN system demonstrated that the system was able to estimate the student's mental states in an accuracy of 76.17%, even if the data recorded by the biometric devices were not normalized.

As future work, it is necessary to further verify the internal consistency and validity of the learner's mental state reports. It is also important to check whether the constructed network has over-learned the training data. In addition, it is necessary to evaluate the practicability of the system by applying the system, which was constructed using experimental data gathered from one learner and one teacher, to other learner-teacher sets.

References

1. Takehana, K., Matsui, T.: Association rules on relationships between learner's physiological information and mental states during learning process. In: Yamamoto, S. (ed.) HIMI 2016. LNCS, vol. 9735, pp. 209–219. Springer, Cham (2016). https://doi.org/10.1007/978-3-319-40397-7_21
2. Fujiyoshi, H., Yoshimura, K., Kunze, K., Kise, K.: Eibun mondai kaitoji no shiten joho o mochiita eigo noryoku suiteho [English ability estimation method using eye movement information during English question answering]. Tech. Rep. Inst. Electron. Inf. Commun. Eng. **115**(25), 49–54 (2015)
3. Horiguchi, Y., Kojima, K., Matsui, T.: A method to estimate learners' impasses based on features in low-level interactions by using MRA. In: Proceedings of the 58th SIG on Advanced Learning Science and Technology, pp. 1–6 (2010)
4. Kojima, K., Muramatsu, K., Matsui, T.: Experimental study on description of eye-movements among choices in answering to multiple-choice problems. Trans. Jpn. Soc. Inf. Syst. Educ. **31**(2), 197–202 (2014)
5. D'Mello, S., Graesser, A., Picard, R.W.: Toward an affect-sensitive AutoTutor. IEEE Intell. Educ. Syst. **22**(4), 53–61 (2007)

6. Matsui, T., Horiguchi, Y., Kojima, K., Akakura, T.: A study on exploration of relationships between behaviors and mental states of learners for value co-creative education and learning environment. In: Yamamoto, S. (ed.) HCI 2014. LNCS, vol. 8522, pp. 69–79. Springer, Cham (2014). https://doi.org/10.1007/978-3-319-07863-2_8
7. Ryu, H., Monk, A.: Analysing interaction problems with cyclic interaction theory: low-level interaction walkthrough. PsychNology J. **2**(3), 304–330 (2004)
8. Fujie, Y.: Role of teacher's repetition in classroom teaching. Jpn. J. Educ. Technol. **23**(4), 201–212 (2000)
9. Kishi, T., Nojima, E.: A structural analysis of elementary school teachers' and children's utterances in Japanese classes. Jpn. J. Educ. Psychol. **54**(3), 322–333 (2006)
10. Shimizu, Y., Uchida, N.: How do children adapt to classroom discourse? Quantitative and qualitative analyses of first grade homeroom activities. Jpn. J. Educ. Psychol. **49**(3), 314–325 (2001)
11. Pekrun, R., Goetz, T., Frenzel, A.C., Barchfeld, P., Perry, R.P.: Measuring emotions in students' learning and performance: the achievement emotions questionnaire (AEQ). Contemp. Educ. Psychol. **36**(1), 36–48 (2011)
12. Tawatsuji, Y., Uno, T., Fang, S., Matsui, T.: Real-time estimation of learners' mental states from learners' physiological information using deep learning, In: Yang, J.C., et al. (eds.) Proceedings of the 26th International Conference on Computers in Education, pp. 107–109 (2018)

Data-Driven Student Clusters Based on Online Learning Behavior in a Flipped Classroom with an Intelligent Tutoring System

Ines Šarić[1], Ani Grubišić[1(✉)], Ljiljana Šerić[2], and Timothy J. Robinson[3]

[1] Faculty of Science, University of Split, Split, Croatia
{ines.saric,ani.grubisic}@pmfst.hr

[2] Faculty of Electrical Engineering, Mechanical Engineering and Naval Architecture, University of Split, Split, Croatia
ljiljana.seric@fesb.hr

[3] Department of Statistics, University of Wyoming, Laramie, USA
tjrobin@uwyo.edu

Abstract. The idea of clustering students according to their online learning behavior has the potential of providing more adaptive scaffolding by the intelligent tutoring system itself or by a human teacher. With the aim of identifying groups of students who would benefit from the same intervention, in this paper, we study a set of 104 weekly behaviors observed for 26 students in a blended learning environment with AC-ware Tutor, an ontology-based intelligent tutoring system. Online learning behavior in AC-ware Tutor is described using 8 tracking variables: (i) the total number of content pages seen in the learning process; (ii) the total number of concepts seen in the learning process; (iii) the total content proficiency score gained online; (iv) the total time spent online; (v) the total number of student logins to AC-ware Tutor; (vi) the stereotype value after the initial test in AC-ware Tutor, (vii) the final stereotype value in the learning process, and (viii) the mean stereotype variability in the learning process. The previous measures are used in a four-step analysis process that includes the following elements: data preprocessing (Z-score normalization), dimensionality reduction (Principal component analysis), the clustering (K-means), and the analysis of a posttest performance on a content proficiency exam. By using the Euclidean distance in K-means clustering, we identified 4 distinct online learning behavior clusters, which we designate by the following names: Engaged Pre-knowers, Pre-knowers Non-finishers, Hard-workers, and Non-engagers. The posttest proficiency exam scores were compared among the aforementioned clusters using the Mann-Whitney U test.

Keywords: Intelligent tutoring system · Blended learning environment · Clustering

A. Coy et al. (Eds.): ITS 2019, LNCS 11528, pp. 72–81, 2019.
https://doi.org/10.1007/978-3-030-22244-4_10

1 Introduction

Feedback is an essential part of education since it helps students raise their awareness of personal strengths and areas for improvement and helps to identify actions for improving their performance. Researchers have revealed that broad-based and personalized interventions can change students' course of learning [1, 2]. A major limitation to the development of classroom-wide interventions is that students' characteristics are highly variable, making it difficult to identify the right intervention. On the other hand, personalized interventions can be time-consuming for teachers, especially when used in large classes. An approach that researchers have used to address these challenges is to identify groups of students who could potentially benefit from the same intervention. The idea of clustering students according to their behavior has the potential of providing more adaptive scaffolding by the system itself (for example in an agent-based intelligent tutoring system) or by a human teacher [3, 4]. There are several research studies that have clustered students into meaningful groups with the goal of informing student interventions [2, 3, 5–7]. Each of these research studies deals with the specific learning environment and the clustering (analysis) approach. Below, we summarize the approaches and findings of several of these studies.

Mojarad et al. [2] investigated data-driven student profiling in a Web-based, adaptive assessment and learning system (ALEKS). The study grouped students into a set of clusters using data from the first half of the semester and 6 key characteristics: the initial assessment score percentage, the total number of assessments, the average days between assessments, the number of days since the initial assessment was taken, an average percentage score increase between assessments, and students' final assessment score percentage in ALEKS (taken at the end of the class). By using Mean shift and K-means clustering algorithms, 5 distinct profiles were identified: Strugglers, Average Students, Sprinters, Gritty, and Coasters. The researchers found these profiles to be useful in enabling institutions and teachers to identify students in need and for subsequently devising and implementing appropriate interventions for groups of students with similar characteristics.

Ferguson and Clow [6] examined the engagement patterns in 4 massive open online courses on a digital education platform (FutureLearn). By using the Silhouette method and the K-means algorithm, the study revealed 7 distinct patterns of engagement: Samplers, Strong Starters, Returners, Mid-way Dropouts, Nearly There, Late Completers and Keen Completers. Results were compared with an earlier study conducted by Kizilcec et al. [8] who used massive learning environments and it was demonstrated that patterns of engagement in these environments were influenced by decisions about pedagogy.

Bouchet et al. [3] clustered students according to their interactions with an intelligent tutoring system designed to foster self-regulated learning (MetaTutor). By using an Expectation-Maximization clustering algorithm and 12 student behavior measures, the analysis revealed 3 distinct student clusters. The study also showed there are variations between clusters regarding prompts they received by the system to perform self-regulated learning processes.

Amershi and Conati [5] proposed a data-based user modeling framework for two specific learning environments (the AIspace Constraint Satisfaction Problem (CSP) Applet and the Adaptive Coach for Exploration (ACE) learning environment) and two data sources (logged interface and eye-tracking data). In this framework, the clustering method, an unsupervised approach, was used to initially identify student behavior, while supervised classifiers were used to recognize student behavior categories. The researchers arbitrarily assumed the number of clusters as k = 2 and 3, because they "expected to find a few distinct clusters with their small sample size" [5].

With an aim to determine whether it was possible to identify distinct student groups based on interaction logs alone, Rodrigo et al. [7] used unsupervised clustering in a learning environment with an intelligent tutoring system (Aplusix). The researchers arbitrarily assumed the number of clusters to be k = 2, and by using K-means clustering algorithm they revealed 2 student clusters associated with differing higher-level behaviors and affective states.

All the above-mentioned student clustering studies are exclusively related to online learning and self-regulated learning using intelligent tutor systems. In this research study, we specifically focus on a blended learning environment that combines face-to-face environment with the teacher and online intelligent tutoring system that students use according to preferred pace, time, and a location. The combination of traditional learning and online learning gives students time to reflect, empowering every student to participate, and enables teacher oversight and feedback anytime and anywhere. In our research study with an introductory computer programming course, the blended learning experience uses a flipped classroom environment. In the used flipped classroom environment, the main concepts of traditional in-class lectures are delivered outside of the class, whereas in-class time is used for activities that allow students to engage with content viewed outside of and before class at a deeper cognitive level. Along with face-to-face lectures and laboratory exercises, students use an ontology-based intelligent tutoring system, AC-ware Tutor.

In the following "Data" section, we describe the AC-ware Tutor tutoring system, the research study protocol, and tracking variables that are used to describe online learning behavior. The "Methodology" section describes the clustering techniques used in the analysis, while the results and conclusion are presented in the last two sections.

2 Data

2.1 AC-Ware Tutor

AC-ware Tutor [9] is an ontology-based intelligent tutoring system that is focused on automatic and dynamic generation, and adaptive selection, sequencing, and presentation of course materials. AC-ware Tutor can be adapted for any subject matter, but that subject matter must be 'fed' into the software by the teacher who prepares a concept map for the course material. Therefore, AC-ware Tutor is not domain specific and can be easily used with various knowledge bases. During tutoring in AC-ware Tutor, the domain knowledge is presented in a textual form using a "fed" concept map (a network of nodes and named relations between them). In AC-ware Tutor, the current level of

student's knowledge is represented using stereotypes that determine the complexity, structure, and presentation of course objects used for student's learning and testing. The iterative process of learning and testing is repeated until the learner gains a certain (maximum) knowledge level.

2.2 Research Study

The proposed research study included 53 undergraduate students from the Faculty of Science at the University of Split enrolled in an Introduction to computer programming course 2017/2018. During a 4-week period (before the course's midterm) in a flipped classroom blended learning environment, students were taught a variety of course units: Basic programming concepts; Variables, types, and operators; Algorithmic structures; and Modules and functions. On average, a weekly concept map prepared by the teacher for the purposes of this research study contained 29 concepts, 33 relations, and a total online score of 133 points.

Prior to face-to-face lectures and laboratory exercises that were held for 4 h per week, online instruction using AC-ware Tutor occurred at students' own pace, time and location. Every week, students first used AC-ware Tutor for learning conceptual knowledge that was taught during regular classes of the following week (flipped classroom). The purpose of pre-class learning is to have students better prepared for the class and thus to allow the teacher to spend face-to-face class for clarifying and applying the conceptual knowledge. One week before the study, the entire class was introduced to the notion of concept mapping. For this introduction to concept mapping, the teacher gave the lecture in a traditional way but also included visually bolded key concepts as well as the relations between concepts. At the end of that lecture, the complete concept map of the week's material was provided to the students.

With an aim to observe and assess weekly learning, paper-based posttests were used for determining content proficiency. These weekly concept map posttests were scored on a scale between 0 and 100.

To analyze weekly online learning behavior, we observed 26 students who participated in this study for all 4 weeks. The dataset included 104 student records (4 records per student) representing weekly online learning behavior. Each record consists of a set of knowledge tracking variables [10] and stereotype tracking variables. These variables were designed for AC-ware Tutor, but they can be appropriately applied to other learning environments and systems.

The online learning behavior measures include the following tracking variables: (i) the total number of content pages seen in the learning process (Objects); (ii) the total number of concepts seen in the learning process (Concepts); (iii) the total score gained in AC-ware Tutor (Score); (iv) the total time spent in AC-ware Tutor (Time); (v) the total number of student logins to AC-ware Tutor (Logins); (vi) the stereotype value after the initial test in AC-ware Tutor (StereoAfterIT), (vii) the final stereotype value in the learning process (StereoFinal), and (viii) the mean stereotype variability through the learning process in AC-ware Tutor (StereoMV). All tracking variables are calculated using online learning data logs from AC-ware Tutor.

In this study we are interested in answering the following questions: Can student clusters be identified according to student interaction with AC-ware Tutor? If clusters can be identified, what are the characteristics that distinguish students belonging to different clusters, and in particular, how do these characteristics relate to posttest proficiency exam performance?

3 Methodology

To investigate student groups within a blended environment with the AC-ware Tutor, we adopt a four-step analysis process. The clustering and statistical methods found in the research study by Mojarad et al. [2] are complemented with the nonparametric test to measure the significance of the difference in student performance between revealed clusters. The complete analysis process consists of the following elements: data pre-processing (Z-score normalization), dimensionality reduction (Principal component analysis), the clustering (K-means), and the analysis of a posttest performance on a content proficiency exam.

After data preprocessing using Z-score normalization, the Principal Component Analysis (PCA) is used to check and reduce the online learning behavior tracking variables.

Next, the number of clusters in the dataset is determined. In addition to Mean-Shift Clustering method [11], the number of clusters is determined using the Elbow method, and the Silhouette method. All three methods are used in conjunction with one another, and when the number of clusters is revealed it is used as the input for the clustering algorithm to get the actual groups.

Due to its wide use and high interpretability, K-means clustering algorithm is used to find student groups with similar online learning behavior tracking variables.

After student clusters are revealed, we use statistical methods to compare tracking variables across clusters. Since nonparametric statistical methods do not assume any specific data distribution, the Mann-Whitney U test is used to determine statistical significance.

In addition to online student behaviors, the research study observed weekly concept map paper-based posttests. Therefore, we investigated the difference in posttest results between different clusters. By using the Mann-Whitney U test, we measure the significance of the difference in student performance between revealed clusters. For each analysis in this four-step process, we use the Python programming language (sklearn and scipy libraries).

4 Results

4.1 Data Preprocessing

The result of standardization (or Z-score normalization) is that the variables are rescaled so that they have the properties of a standard normal distribution with $\mu = 0$ and $\sigma = 1$, where μ is the mean (average) and σ is the standard deviation from the mean; standard scores (also called z scores) of the samples are calculated as follows: $z = (x - \mu)/\sigma$ where x denotes the score on the tracking variable. Standardizing the features so that they are centered around 0 with a standard deviation of 1 is important because we are comparing measurements that have different units, and it is also a general requirement for many machine learning algorithms, such as PCA.

4.2 Dimensionality Reduction

PCA is a statistical procedure that uses an orthogonal transformation to convert a set of observations of possibly correlated variables into a smaller number of linearly uncorrelated variables known as "principal components" (PCs). For our dataset, PCA revealed 5 PCs from Table 1 that cumulatively explain nearly 92.6% of the variance in the data. Therefore, we use these PCs as the input to Mean shift clustering and K-means algorithm. It is important to note that used PCs refer strongly to all online learning behavior variables, and for that reason, the selection of variables is additionally confirmed.

Table 1. Tracking variables' weights and explained variance proportion for PCs.

PC	1	2	3	4	5
Objects	0.354	−0.346	−0.249	−0.199	0.706
Concepts	−0.237	−0.394	0.25	−0.779	−0.258
Score	−0.259	−0.518	−0.123	0.034	0.17
Time	0.273	−0.506	0.024	0.333	−0.529
Logins	0.274	−0.157	0.868	0.201	0.247
StereoAfterIT	−0.459	0.19	0.277	0.003	0.194
StereoFinal	−0.377	−0.369	−0.134	0.404	0.076
StereoMV	0.499	0.049	−0.106	−0.196	−0.146
Explained variance proportion	0.456	0.236	0.102	0.084	0.048

4.3 Clustering

Mean shift clustering detects the modes of the density using kernel density estimation, and the number of modes can be used to decide the number of clusters in data by finding the centers of mass within the data [2]. In addition to Mean shift clustering, we also check the result of the Elbow and the Silhouette method. When all methods are taken into account, the number of clusters in our study is k = 4. In case of Mean shift clustering, the size of each cluster was the decisive factor. When looking at the Elbow method plot, the curve is visible for k = 4 clusters. Additionally, the appropriate value of the average Silhouette score for k = 4 confirmed the previous results. Therefore, 4 possible student clusters are identified in the research study data.

For k = 4 clusters, we use K-means clustering and the Euclidean distance as the distance measure to look for 4 distinct groups of students, using the data from the first 5 PCs. To interpret these clusters, we then look at the average values for each variable in each cluster and determine whether each cluster has very low, low, medium or high average values for each variable, in relation to other clusters, shown in Table 2.

Table 2. Average values of online learning behavior tracking variables for each cluster.

Cluster	1 Non-engagers		2 Pre-knowers Non-finishers		3 Hard-workers		4 Engaged Pre-knowers	
Students	16		31		24		33	
Objects	2.563	(Average)	1.516	(Low)	4.375	(High)	1.758	(Low)
Concepts	**25.375**	**(High)**	26.677	(High)	**27.833**	**(High)**	**32.03**	**(High)**
Score	**44.438**	**(Low)**	**93.774**	**(Average)**	**119.625**	**(High)**	**147.424**	**(High)**
Time	21.123	(Low)	9.969	(Very low)	**53.551**	**(High)**	16.589	(Low)
Logins	2.063	(High)	1.065	(Average)	1.875	(High)	1.182	(Average)
StereoAfterIT	**1.438**	**(Low)**	**3.548**	**(High)**	**0.792**	**(Very low)**	**3.545**	**(High)**
StereoFinal	1.625	(Low)	3.968	(High)	**3.792**	**(High)**	**3.97**	**(High)**
StereoMV	2.566	(High)	0.242	(Very low)	1.625	(Average)	0.207	(Very low)

The revealed clusters can be interpreted using characteristics such as the prior knowledge, the online engagement, and/or the completion of online learning assignments. The prior knowledge is described by the StereoAfterIT tracking variable, while the online engagement is described by the combination of tracking variables – the Concepts, the Time, the Score and the StereoFinal. The completion of online learning assignments is described by the combination of the Concepts and the Score variables.

By using the cluster tracking variables from Table 2, we have identified 4 types of AC-ware Tutor students. Each cluster is named according to its comparative differences. Engaged Pre-knowers are students with high prior knowledge (the StereoAfterIT) and high online engagement (the Concepts, the Score and the StereoFinal) (Cluster 4). For Engaged Pre-knowers, it took less time to complete the online tutoring than for the other student groups and Engaged Pre-knowers had very low variability in stereotype value. Pre-knowers Non-finishers are students with also high prior knowledge (the StereoAfterIT), but these students did not complete their online assignments (the Score) (Cluster 2). Pre-knowers Non-finishers spent the least amount of time online and despite seeing most of the concepts in the learning process they did not complete their online learning assignments. Hard-workers are students with very low prior knowledge (the StereoAfterIT) and high online engagement (the Concepts, the Score, the Time and the StereoFinal) (Cluster 3). Since Hard-workers spent the most time online, these students learned from the highest number of content pages. The Non-engagers are students with low prior knowledge (the StereoAfterIT) who did not complete their online learning assignments (the Concepts and the Score) (Cluster 1). The previous students had on average the highest variability in stereotype value. In terms of the average time spent online, Engaged Pre-knowers spent online 16 min, Pre-knowers Non-finishers 9 min, Hard-workers 53 min, and Non-engagers 21 min.

Since nonparametric statistical methods do not assume any specific data distribution, we use the Mann-Whitney U test to determine if clusters are statistically different in terms of each attribute. Table 3 shows the statistical significance (p-value) of the difference between each attribute between each pair of clusters. As Table 3 shows, each pair of clusters is different in at least 6 of 8 variables, except Cluster 2 and Cluster 4 that differ in 3 variables (Concepts, Score and Time). The previous finding suggests that proposed online learning behavior tracking variables can be used to define and describe 4 student clusters and that Pre-knowers Non-finishers (Cluster 2) and Engaged Pre-knowers (Cluster 4) can be further analyzed in future research.

Table 3. Statistical significance (p-value) of the difference between each pair of clusters.

Clst1	Clst2	Objects	Concepts	Score	Time	Logins	StereoAfterIT	StereoFinal	StereoMV
1	2	**p = 0.015**	p = 0.406	**p = 0.000**	**p = 0.018**	**p = 0.000**	**p = 0.000**	**p = 0.000**	**p = 0.000**
1	3	**p = 0.003**	p = 0.057	**p = 0.000**	**p = 0.000**	p = 0.336	**p = 0.025**	**p = 0.000**	**p = 0.000**
1	4	p = 0.078	**p = 0.000**	**p = 0.000**	p = 0.347	**p = 0.000**	**p = 0.000**	**p = 0.000**	**p = 0.000**
2	3	**p = 0.000**	**p = 0.027**	**p = 0.004**	**p = 0.000**	**p = 0.000**	**p = 0.000**	**p = 0.044**	**p = 0.000**
2	4	p = 0.068	**p = 0.000**	**p = 0.000**	**p = 0.004**	p = 0.081	p = 0.354	p = 0.491	p = 0.411
3	4	**p = 0.000**	**p = 0.000**	**p = 0.000**	**p = 0.000**	**p = 0.000**	**p = 0.000**	**p = 0.038**	**p = 0.000**

4.4 Clusters and Mastery

We have checked if there is a difference between revealed clusters according to student paper-based posttest performance. Figure 1 contains boxplots for the posttest scores of each student cluster. Based on their online learning behavior, our hypothesis is that Engaged Pre-knowers, Pre-knowers Non-finishers, and Hard-workers perform better than Non-engagers, what was confirmed after using Mann-Whitney U test.

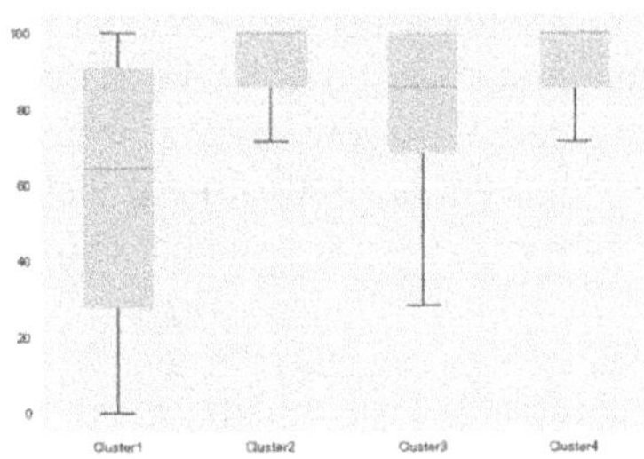

Fig. 1. The boxplot visualization for the posttest scores of Cluster 1, 2, 3 and 4.

The mean values and standard deviations (SD) for each cluster are shown in Table 4. As we can see from boxplot visualizations, the groups differ in their mastery according to the stated hypothesis. By using the Mann-Whitney U test, we measure the significance of the difference in student performance between revealed clusters. The results of the Mann-Whitney U test are shown in Table 5, where we can see that 4 out of 6 analysis confirmed the statistical significance of different clusters' performances. There is a statistically significant difference in posttest performance between Cluster 1 and 2 (Non-engagers and Pre-knowers Non-finishers) and Cluster 1 and 4 (Non-engagers and Engaged Pre-knowers). We do not find statistical significance in posttest performance between Clusters 1 and 3 (Non-engagers and Hard-workers).

Table 4. Statistical significance of differences between clusters' performances.

Lbl	Cluster name	Mean	SD
1	Non-engagers	58.506	34.580
2	Pre-knowers non-finishers	90.290	16.863
3	Hard-workers	72.690	31.327
4	Engaged pre-knowers	90.869	15.214

Table 5. Statistical significance of differences between clusters' performances.

Cluster 1	Cluster 2	P-value
1	2	**p = 0.000**
1	3	p = 0.139
1	4	**p = 0.000**
2	3	**p = 0.006**
2	4	p = 0.500
3	4	**p = 0.004**

5 Conclusion

In this study, we aimed to address the challenge of finding student clusters based on online learning behavior in AC-ware Tutor. In the case of identifying online learning patterns, students of each cluster could potentially benefit from the same intervention. We examined the flipped classroom submodel of a blended learning environment in which students learn online in advance. We used cluster analysis techniques to identify groups based on eight online learning behavior measures. By using the Euclidean distance in K-means clustering, we have identified 4 distinct online learning behavior clusters: Engaged Pre-knowers, Pre-knowers Non-finishers, Hard-workers, and Non-engagers. Since nonparametric statistical methods do not assume any specific data distribution, we have used the Mann-Whitney U test to measure the statistical significance of the difference in each attribute between each pair of clusters. It is revealed that each pair of clusters is different in at least 6 of 8 variables, except Cluster 2 and 4 (Pre-knowers Non-finishers, Engaged Pre-knowers) that differ in 3 variables (Concepts, Score and Time). Identified clusters have also been analyzed whether they differ in terms of students' posttest performances. The student clusters differ in their mastery in the hypothesized fashion, in which Engaged Pre-knowers and Pre-knowers Non-finishers performed better than Non-engagers. When we checked the posttest performances with the Mann-Whitney U test, the difference could not be confirmed between Clusters 1 and 3 (Non-engagers and Hard-workers).

The previous clusters can be used by teachers to devise appropriate interventions in time to still take meaningful action – at the course's midterm. For example, the Engaged Pre-knowers and Pre-knowers Non-finishers can be motivated with additional assignments to maintain learning consistency, while Non-engagers can be warned out to put higher efforts in the learning process. Perhaps it would be helpful to let students know that they will fall into one of four group types and that they can be most successful if they approach the class by having fewer logins but more time spent on each login. Also, for students that have very little prior knowledge as evidenced by the pretest, they will need to have a higher engagement in terms of logins and time spent during each login. Perhaps there can be a dashboard in an online system that students can use that shows their number of logins along with time spent in each login as compared to the rest of the class - if they find themselves to be low in terms of engagement, they will know that they should pick up their effort.

Beside human teacher's interventions, the further research should investigate possibilities of system's interventions. The next step should also be to address some of the study and analysis limitations, such as to collect more data to strengthen the results and to explore alternative clustering techniques. Successful identification of student clusters is a pre-requisite for more effective and adaptive learning delivered by a human teacher or intelligent tutoring system.

Acknowledgement. This paper is part of the Adaptive Courseware & Natural Language Tutor project that is supported by the Office of Naval Research Grant No. N00014-15-1-2789.

References

1. Lin-Siegler, X., Dweck, C.S., Cohen, G.L.: Instructional interventions that motivate classroom learning. J. Educ. Psychol. **108**, 295–299 (2016)
2. Mojarad, S., Essa, A., Mojarad, S., Baker, R.S.: Data-driven learner profiling based on clustering student behaviors: learning consistency, pace and effort. In: Nkambou, R., Azevedo, R., Vassileva, J. (eds.) ITS 2018. LNCS, vol. 10858, pp. 130–139. Springer, Cham (2018). https://doi.org/10.1007/978-3-319-91464-0_13
3. Bouchet, F., Harley, J.M., Trevors, G.J., Azevedo, R.: Clustering and profiling students according to their interactions with an intelligent tutoring system fostering self-regulated learning. J. Educ. Data Min. JEDM. **5**, 104–146 (2013)
4. Vellido, A., Castro, F., Nebot, À.: Clustering educational data. In: Romero, C., Ventura, S., Pechenizkiy, M., Baker, R. (eds.) Handbook of Educational Data Mining, pp. 75–92. CRC Press (2010)
5. Amershi, S., Conati, C.: Combining unsupervised and supervised classification to build user models for exploratory. J. Educ. Data Min. JEDM. **1**, 18–71 (2010)
6. Ferguson, R., Clow, D.: Examining engagement: analysing learner subpopulations in massive open online courses (MOOCs). In: Proceedings of the 5th International Conference on Learning Analytics and Knowledge - LAK 2015, pp. 51–58. ACM, Poughkeepsie (2015)
7. Rodrigo, M.M.T., Angloa, E.A., Sugaya, J.O., Baker, R.S.J.D.: Use of unsupervised clustering to characterize learner behaviors and affective states while using an intelligent tutoring system. In: International Conference on Computers in Education (2008)
8. Kizilcec, R.F., Piech, C., Schneider, E.: Deconstructing disengagement: analyzing learner subpopulations in massive open online courses. In: Proceedings of the 3rd International Conference on Learning Analytics and Knowledge - LAK 2013, pp. 170–179. ACM, New York (2013)
9. Grubišić, A.: Adaptive student's knowledge acquisition model in e-learning systems, Ph.D. thesis, University of Zagreb, Croatia (2012)
10. Grubišić, A., et al.: Knowledge tracking variables in intelligent tutoring systems. In: Proceedings of the 9th International Conference on Computer Supported Education - CSEDU 2017, pp. 513–518. SCITEPRESS, Porto (2017)
11. Arnold, K.E., Pistilli, M.D.: Course signals at Purdue: using learning analytics to increase student success. In: Proceedings of the 2nd International Conference on Learning Analytics and Knowledge - LAK 2012, pp. 267–270. ACM, New York (2012)

Decision Support for an Adversarial Game Environment Using Automatic Hint Generation

Steven Moore(✉) and John Stamper

HCII, Carnegie Mellon University, Pittsburgh, USA
StevenJamesMoore@gmail.com, jstamper@cs.cmu.edu

Abstract. The Hint Factory is a method of automatic hint generation that has been used to augment hints in a number of educational systems. Although the previous implementations were done in domains with largely deterministic environments, the methods are inherently useful in stochastic environments with uncertainty. In this work, we explore the game Connect Four as a simple domain to give decision support under uncertainty. We speculate how the implementation created could be extended to other domains including simulated learning environments and advanced navigational tasks.

Keywords: Hint generation · Educational data mining · Reinforcement learning

1 Introduction and Related Work

Adaptive learning through the use of intelligent tutoring systems (ITS) have been shown to increase overall learning and decrease the time needed to mastery (Koedinger et al. 2013). To overcome the difficulty of ITS creation, developers have turned to authoring tools, like CTAT (Aleven et al. 2006) or data driven techniques (Stamper et al. 2007). While large educational data repositories like DataShop (Stamper et al. 2010) exist for many traditional educational systems, one area that has been less explored is decision support systems in domains that include multiple interacting students (or players or agents). These domains are significantly more complex because, unlike a traditional educational system, a specific action in these domains by a student may not have a deterministic outcome. In this work, we explore how a technique called the Hint Factory (Stamper et al. 2008) can be used for decision support (hints and feedback) in adversarial domains and specifically implement the technique using a Connect Four data set.

Previous work has utilized datasets from solved two-player games, such as Connect Four, often for reinforcement learning (RL) and temporal difference learning (TDL) tasks. One such study found that using a self-learning agent operating with the Minimax algorithm required 1.5 million games to establish an 80% success rate (Thill et al. 2012). We look to show how well the Hint Factory can be utilized in such domains, like adversarial games, in order to provide decision support in the form of hints and feedback.

A. Coy et al. (Eds.): ITS 2019, LNCS 11528, pp. 82–88, 2019.
https://doi.org/10.1007/978-3-030-22244-4_11

The Hint Factory provides context specific hints to students using a novel technique that creates graphs of past student solutions from student log data, which can then be used to suggest the best step for a student to solve the problem based on their current state in the graph. The Hint Factory, applied in a tutor for teaching deductive logic proofs, has been shown to significantly increase learning and reduce attrition (Stamper et al. 2011). Determining the timing and frequency of hints is a particular challenge, but studies suggest that offering hints on demand, instead of proactively, can have positive effects on learning (Razzaq and Heffernan 2010). While some studies have suggested as much as 72% of help-seeking behaviors can be unproductive (Aleven et al. 2004), Shih's work suggests that some of these behaviors are in fact helpful. They argue that using help to achieve a bottom-out hint can be seen as looking for a worked example, an effective learning strategy (Shih et al. 2011). On-demand, context-specific Hint Factory hints and feedback are effective in single user educational technologies and have been shown to significantly help students complete more problems resulting in higher outcomes on post tests (Stamper et al. 2011).

In Connect Four, there are three possible states for each of the forty-two available game spaces, with a game board configuration consisting of six rows and seven columns. The space can be occupied by the turn players piece, the opponent's piece, or it can be empty. Early work by Allis (1988) found that after ruling out possible illegal configurations that were present in the initial estimates, the possible board configurations were close to an upper bound of $7.1 * 10^{13}$. More recently, it was determined using binary decision diagrams that there are exactly 4,531,985,219,092 legal board configurations (Edelkamp and Kissmann 2008).

A key feature of the Hint Factory is the use of a Markov Decision Process (MDP) to create a policy based on data from previous attempts at problems in an educational system (Stamper 2006). The resulting policy allows the Hint Factory to determine the next best state and action a student can take. The features of the state and action are used to generate hints and feedback for the student or user (Barnes et al. 2011). In this work, states are represented by the current configuration of the Connect Four board at a given time step and a time step is the state of the board after both players have made their move. Actions cause the current state to transition into a new state, for this game there are at most seven possible actions represented by the column a game piece can be dropped into. Transitions, which have an associated probability, is the process of entering a specific new state, based on the current state and the player's chosen action. In Connect Four, a given state-action pair can result in at most seven new states, without accounting for the following adversarial agent response, which also moves into one of the seven available columns. This means that at most one state can lead into forty-nine following states accounting for both players' actions.

Rewards are a numerical value tied to entering a new state, we assign a high value for a winning endgame state, and a high negative value for a losing endgame state, and 0 for an endgame state resulting in a draw. All other game states begin with a value of 0 and then value iteration is applied to create a policy for the MDP (Sutton and Barto 1998). Our implementation utilizes a greedy policy, which prioritizes being on a path toward any winning state, regardless of how many turns it took. Finally, the value is the expected reward starting from a state, taking an action, and then following the given policy in place.

2 Implementation

To collect the Connect Four data, we used a web-based implementation with the logic of the game written in TypeScript (Anggara 2018). The program puts two computer players against one another to simulate the gameplay, recording each of their moves. In total we collected five thousand games played between the two computer players.

Both computer players used the same logic to play the game, which is the Minimax algorithm with alpha-beta pruning. The Minimax algorithm looks at all possible states, determines the worst possible outcome, and then selects a possible state that has the least worst outcomes (Edwards 1954). One non-standard modification we did to the algorithm is that the first move for each computer player is randomized. We found that causing each player's first move to randomly be placed in one of the seven columns created a greater diversity of games that closely resembled gameplay between two human players. Additionally, Minimax uses a heuristic function that weighs a reward in the closer future worth more than the same reward in the distant future. This means it favors moves that win the next turn, prevent the opponent from winning, and gravitate toward the center of the game board, as a human player would.

Using Python, we implemented a version of the Hint Factory for Connect Four. Our implementation follows a greedy policy that prioritizes the highest immediate reward, regardless of what the future rewards might be. While this may seem like an impatient policy, the adversarial and stochastic nature of Connect Four makes policies relying on potential future rewards more risky. With our aforementioned policy, the value for the states is the reward of the greatest state-action pairing. It indicates which action to select, based on the calculated rewards with transition probabilities, that leads to a state most commonly found on the path to a winning game.

3 Results and Discussion

Our dataset consists of 5,000 Connect Four games, with an average of twenty-six turns (13 states) per game. If we use that average and process all 5,000 games, it results to 65,000 states. However, when we process all 65,000 states for the games, only 2,606 (4.15%) of that total are unique game states. Table 1 summarizes the unique states encountered at different processing increments of the total games.

Table 1. Number of unique states generated from games played in a simulated study.

Games processed	Unique states	Percent increase
100	797	–
500	1,531	92.10
1,000	1,865	21.82
2,000	2,197	17.80
3,000	2,386	8.60
4,000	2,510	5.20
5,000	2,606	3.82

As noted, the percentage of new states encountered continues to decrease, as expected, as we introduce more games. While there are over 4.5 trillion legal board states, many of them will not be reached if players are following general strategies of winning. For instance, human players might follow the strategy of trying to maximize their game pieces in the center columns, causing many of the states to revolve around a similar configuration. This is the case for our agents, both of which follow the same Minimax algorithm and heuristic function in attempts to win. The observed games begin randomly, but the agent's moves tend to gravitate toward the center as they follow the heuristic for winning, much like a human player might.

Using the first 1,000 games to build the model, we were able to provide hints for roughly 50,000 of the 52,000 states that were present in the following 4,000 games. The 52,000 states making up the later 4,000 games includes repeated states, as some of the games followed the same sequence. For our 5,000 observed games, 1,123 of them were unique. This reduces the space from 65,000 total states to 14,600 states if we only include unique games. Building the model based off the first 1000 random games recorded yields 1,865 unique states for our system. Using that data, we were able to provide a hint, suggesting the next optimal move to the user, for all but 1,931 of the encountered states in the remaining 4,000 games. That results in this implementation being able to provide a hint 87.23% of the time when using just the first 20% of the recorded data.

Next-step hints, such as the ones generated by our system, are often not as descriptive as instructor generated ones, causing some educators to doubt their effectiveness for learning. However, such hints have been proven effective, as in a study by Rivers (2017) which resulted in students spending 13.7% less time on practice while achieving the same learning results. Paquette et al. (2012) found that next-step hints yielded increased learning gains within an ITS, compared to offering only flag feedback, and that they were as efficient and appreciated as instructor generated hints. In stochastic learning environments that often lack hints altogether, generating next-step hints using Hint Factory techniques can improve educational effectiveness and easily integrates into an ITS or other educational system.

Increasing complexity is brought on from adversarial agents when a policy is being determined for the system. An adversarial agent creates uncertainty on what the state will be, even when the system suggests a particular action for the player. In the case of our Connect Four implementation, it is not truly random as a heuristic is being followed via the Minimax algorithm. We can offer a hint, suggesting where the player should place their game piece, based on what we have previously seen and its success. However, we have no other source that suggests what action the adversarial agent will take, in this case the column into which they will drop their game piece. This type of uncertainty can make certain policies seemingly optimal, but they might not be as successful when different data is utilized to build the model or when the adversarial agent plays more randomly. Such uncertainty has previously made these domains difficult to provide next-step hints for, as the following state remains unknown.

Hints are effective for student learning and have been proven effective in open-ended domains (Stamper et al. 2011; Rivers 2017), yet their data-driven implementation remains difficult in complex stochastic environments such as immersive training simulation environments One instance of this in chemistry is the VLab, a virtual environment for students to conduct experiments and interpret the reactions (Tsovaltzi et al. 2010). If students are working on a stoichiometry problem, the reaction will vary based on how much and what type of a certain chemical they add, thus dictating how much of another substance they need to add next. Detailed hints for a system may be hard to generate because they are dependent on the many combinations of chemical type and quantity that could be added. However, it would be a good fit for our techniques, using either collected or simulated student data, to provide hints.

In physical stochastic environments, such as driving a car or operating a robotic arm, many adversarial forces are at play, such as objects in the path or even gravity. Next-step hints in such domains can provide needed decision support and be generated using similar methods as applying the Hint Factory to Connect Four, without requiring mass amounts of data or computational power. In the realm of robotics, Lego Mindstorms have been a popular and effective mechanism for teaching middle and high school students programming and computational thinking (Mosley and Kline 2006). Such hints could assist students in the debugging of their robots, as they get them to perform a given task, taking into account the physical forces they encounter. Similar techniques could be applied to programmatic physics engines used for student experimentation, which often have many reactive forces at play.

4 Conclusion and Future Work

The stochastic feature of adversarial game environments naturally fit the Hint Factory's underlying MDP mechanisms. The main contribution of this work is to show how the Hint Factory can be used to create a hint and feedback policy in an adversarial game, that can be extended to more complex educational domains. The novel implementation in this stochastic domain shows that with a relatively small selection of the overall state space, we can provide coverage to provide hints and feedback for a large percentage of commonly seen states. This has real implications for many other multiagent and adversarial game and learning environments.

We see some obvious next steps for this work in other multi-agent environments, particularly simulations of the physical world where many forces are at play. Determining what to track as part of the state and what encompass an action will need to be addressed as we move into more complex stochastic domains. Next we plan to execute experiments using the hint generations to see if we get similar gains to the Hint Factory implementations in deterministic domains. We also want to look at how human users learn from the suggested hints and mimic any strategies being conveyed.

References

Aleven, V., McLaren, B., Roll, I., Koedinger, K.: Toward tutoring help seeking. In: Lester, J.C., Vicari, R.M., Paraguaçu, F. (eds.) ITS 2004. LNCS, vol. 3220, pp. 227–239. Springer, Heidelberg (2004). https://doi.org/10.1007/978-3-540-30139-4_22

Aleven, V., McLaren, B.M., Sewall, J., Koedinger, K.R.: The cognitive tutor authoring tools (CTAT): preliminary evaluation of efficiency gains. In: Ikeda, M., Ashley, K.D., Chan, T.-W. (eds.) ITS 2006. LNCS, vol. 4053, pp. 61–70. Springer, Heidelberg (2006). https://doi.org/10.1007/11774303_7

Allis, L.V.: A knowledge-based approach of connect-four. Vrije Universiteit, Subfaculteit Wiskunde en Informatica (1988)

Anggara, K.: Connect Four (2018). https://doi.org/10.5281/zenodo.1254572

Barnes, T., Stamper, J., Croy, M.: Using Markov decision processes for automatic hint generation. In: Handbook of Educational Data Mining, 467 (2011)

Edelkamp, S., Kissmann, P.: Symbolic classification of general two-player games. In: Dengel, A. R., Berns, K., Breuel, T.M., Bomarius, F., Roth-Berghofer, T.R. (eds.) KI 2008. LNCS (LNAI), vol. 5243, pp. 185–192. Springer, Heidelberg (2008). https://doi.org/10.1007/978-3-540-85845-4_23

Edwards, W.: The theory of decision making. Psychol. Bull. **51**(4), 380 (1954)

Koedinger, K.R., Stamper, J.C., McLaughlin, E.A., Nixon, T.: Using data-driven discovery of better student models to improve student learning. In: Lane, H.C., Yacef, K., Mostow, J., Pavlik, P. (eds.) AIED 2013. LNCS (LNAI), vol. 7926, pp. 421–430. Springer, Heidelberg (2013). https://doi.org/10.1007/978-3-642-39112-5_43

Mosley, P., Kline, R.: Engaging students: a framework using lego robotics to teach problem solving. Inf. Technol. Learn. Perform. J. **24**(1), 39–45 (2006)

Paquette, L., Lebeau, J.-F., Beaulieu, G., Mayers, A.: Automating next-step hints generation using ASTUS. In: Cerri, S.A., Clancey, W.J., Papadourakis, G., Panourgia, K. (eds.) ITS 2012. LNCS, vol. 7315, pp. 201–211. Springer, Heidelberg (2012). https://doi.org/10.1007/978-3-642-30950-2_26

Razzaq, L., Heffernan, N.T.: Hints: is it better to give or wait to be asked? In: Aleven, V., Kay, J., Mostow, J. (eds.) ITS 2010. LNCS, vol. 6094, pp. 349–358. Springer, Heidelberg (2010). https://doi.org/10.1007/978-3-642-13388-6_39

Rivers, K.: Automated Data-Driven Hint Generation for Learning Programming (2017)

Shih, B., Koedinger, K.R., Scheines, R.: A response time model for bottom-out hints as worked examples. In: Handbook of Educational Data Mining, pp. 201–212 (2011)

Stamper, J.C., Eagle, M., Barnes, T., Croy, M.: Experimental evaluation of automatic hint generation for a logic tutor. In: Biswas, G., Bull, S., Kay, J., Mitrovic, A. (eds.) AIED 2011. LNCS (LNAI), vol. 6738, pp. 345–352. Springer, Heidelberg (2011a). https://doi.org/10.1007/978-3-642-21869-9_45

Stamper, J., Barnes, T., Croy, M.: Enhancing the automatic generation of hints with expert seeding. Int. J. AI Educ. **21**(1–2), 153–167 (2011b)

Stamper, J., et al.: PSLC DataShop: a data analysis service for the learning science community. In: Aleven, V., Kay, J., Mostow, J. (eds.) ITS 2010. LNCS, vol. 6095, p. 455. Springer, Heidelberg (2010). https://doi.org/10.1007/978-3-642-13437-1_112

Stamper, J., Barnes, T., Lehmann, L., Croy, M.: The hint factory: Automatic generation of contextualized help for existing computer aided instruction. In: 9th International Conf on Intelligent Tutoring Systems Young Researchers Track, pp. 71–78 (2008)

Stamper, J.C., Barnes, T., Croy, M.: Extracting student models for intelligent tutoring systems. In: Proceedings of the National Conference on Artificial Intelligence, vol. 22, no. 2, p. 1900. AAAI Press; MIT Press, 1999 (2007)

Stamper, J.: Automating the generation of production rules for intelligent tutoring systems. In: Proceedings of the 9th International Conference on Interactive Computer Aided Learning (ICL 2006). Kassel University Press (2006)

Sutton, R., Barto, A.: Reinforcement Learning. The MIT Press, Cambridge (1998)

Thill, M., Koch, P., Konen, W.: Reinforcement learning with N-tuples on the game connect-4. In: Coello, C.A.C., Cutello, V., Deb, K., Forrest, S., Nicosia, G., Pavone, M. (eds.) PPSN 2012. LNCS, vol. 7491, pp. 184–194. Springer, Heidelberg (2012). https://doi.org/10.1007/978-3-642-32937-1_19

Tsovaltzi, D., et al.: Extending a virtual chemistry lab with a collaboration script to promote conceptual learning. Int. J. Technol. Enhanc. Learn. **2**(1–2), 91–110 (2010)

Detecting Collaborative Learning Through Emotions: An Investigation Using Facial Expression Recognition

Yugo Hayashi(✉)

College of Comprehensive Psychology, Ritsumeikan University,
2-150 Iwakura-cho, Osaka, Ibaraki 567-8570, Japan
y-hayashi@acm.org

Abstract. Providing adaptive feedback to learners engaging in collaborative learning activities is one research topic in the development of intelligent tutoring systems. However, there is a need to investigate how to systematically evaluate a learner's activities and provide feedback on them. The present study investigates how emotional states, detected through facial recognition, can be utilized to capture the learning process in a simple jigsaw-type collaborative task. It was predicted that when learners argue with each other and reason deeply, they may experience several emotional states such as positive and negative states. The results show that when learners work harder on developing a mutual understanding through conflictive interaction, negative emotions can be used to predict this process. This study contributes to the knowledge of how emotional states detected by facial recognition technology can be applied to predict learning process in conflictive tasks. Moreover, these empirical results will impact the development of adaptive feedback mechanisms for intelligent tutoring systems for collaborative learning.

Keywords: Collaborative learning · Pedagogical conversational agents · Emotion · Learning assessment

1 Introduction

Intelligent tutoring systems have been long investigated in educational science [3,8,9,16,21,28], and one of the goals of these systems is to detect the learners' mental states and adaptively provide feedback. The use pedagogical conversational agents (PCAs) demonstrating benefits in learning gains has emerged in the last decade [8,20,22]. Recently, social learning such as learner-learner collaborative learning has come to be regarded as an important skill for the 21st century, and several studies have used PCAs in the context of collaborative learning. However, in cognitive and learning science, the mechanisms of collaborative interactions and their related process are not fully understood. This paper demonstrates how facial expression recognition can be used to predict emotional states to evaluate collaboration.

A. Coy et al. (Eds.): ITS 2019, LNCS 11528, pp. 89–98, 2019.
https://doi.org/10.1007/978-3-030-22244-4_12

1.1 Collaborative Learning and Intelligent Systems

Studies in cognitive science show that collaboration helps to externalize knowledge [25,27] as well as facilitate meta-cognition during explanations [4] and perspective change [14]. It has been noted that social-conflict-based learning plays an important role in the learning process [19,29], and collaborative learning takes advantage of the nature of such conflict-based learning. Several studies in this area have attempted to understand these activities [7,15]. The 2015 Programme for International Student Assessment, which is administered by the Organisation for Economic Co-operation and Development, has surveyed student skills and knowledge [26]. In these surveys, two skills that leverage collaborative learning, task-work and team-work, are considered to be important skills. The former is related to problem-solving skills and the latter is related to social skills such as the ability to coordinate and establish common ground with other group members.

Although team-work plays an important role in collaborative learning, it has been difficult to quantitatively evaluate a learner's conversational behaviors with respect to the success or failure of team interactions. It is hence a challenge to develop computational models of a learner's interactions to automatically detect his/her state and provide group facilitation accordingly. There have been studies that have successfully detected a learner's state from linguistic data and used a PCA to assist learning [18]. However, it is still difficult to completely understand the detailed context of social interactions. There have also been attempts to use multiple variables such as verbal and nonverbal channels [5]; however, it is not fully understood which paradigm is best for evaluating both team-work and task-work in collaborative learning.

1.2 Using Emotional States as Predictors for Learning

Studies in collaborative learning have used tasks that generate social conflict, such as a jigsaw-type task, in which learners tend to discuss their different perspectives and conflictive states are expected to occur during the task [1]. On such occasions, it is likely that confusion and arguments may occur. As a result, learners may experience emotional states [6]. The study [8] showed that learning gains were positively correlated with confusion and engagement/flow, negatively correlated with boredom, and were uncorrelated with the other emotions. Moreover, psychology studies on general problem solving have discovered that when problem solvers are confronted in an impasse, the emotional states are highly related [24]. According to these studies, positive emotions are highly related to aptitude tasks. Other studies have also shown that positive emotions play an important role in interactive behavior [10].

These studies imply that emotional states, especially emotional states that are related to negative/positive feelings can be used to detecting a learner's performance and role in collaborative learning in a conflictive task. However, it is important to consider that the instances of confusion that are peripheral to the learning activity are unlikely to have any meaningful impact on learning [6].

In the present study, we use a jigsaw-type task that includes the integration of other learners' different perspectives. It is expected that, to establish common ground in order to achieve the task, learners will experience emotional states: a negative state during confusion and conflicts during and a positive state when communication is successful.

1.3 Goal and Hypotheses

The goal of this study is to investigate how an emotional state that is detected during collaborative learning can be used to predict performance in a conflictive collaborative learning task. The long-term goal of this research is to develop an adaptive collaborative tutoring system in which PCAs (developed in the authors' previous work) facilitate learning according to the learners' states.

In this study, we investigate collaborative learning in a jigsaw-type task in which socio-cognitive conflict is expected to occur. It is predicted that, to achieve the task, learners may experience both positive and negative emotions due to the nature of the task. We hence consider the following hypothesis (H1): when experiencing arguments during the task, learners become conflictive and confused, and these can be detected as negative emotions. Hypothesis H1 has two parts: more strongly negative emotions are related to higher-level coordination activities such as establishing common ground about their different knowledges (H1-a) and thus affect learning performance (H1-b). We also consider the following hypothesis (H2): learners experience positive emotions when establishing common ground and reaching agreement (H2-a) and these emotions thus influence learning performance (H2-b).

The present study investigates these hypotheses by focusing on emotions detected using learners' facial expressions. This use of artificial intelligence technology supports our long-term goal of developing intelligent and adaptive tutoring systems.

2 Method

2.1 Procedure and Task

Twenty Japanese university students participated in dyads in this experiment. The participants received course credit for participation. This study was conducted after passing an ethical review conducted by the authors' institutional ethical review committee.

The experiment design consisted of a pre-test, main task, and post-test procedure. The main task of this experiment was to explanation a topic that was taught in a cognitive science class. They were required to explain the phenomenon of how humans process language information and were required to use two sub-concepts: "top-down processing" and "bottom up processing." This study adopts the jigsaw method [1], which is a style of learning in which each learner has knowledge of one of the sub-concepts and exchanges it with their partner through explanation. The learners' goal was to explain their different perspectives and provide an overall integrated explanation of the phenomenon using the two sub-concepts. To achieve their goal, they were required to argue about how each sub-concept can be used to explain the main concept.

Participants individually worked on the pre-test to determine whether they already knew about the sub-concepts. The main task was conducted for ten minutes. After completion, the learners again individually performed the post-test so that their knowledge gain could be measured.

2.2 Experimental Set-Up

The experiment was conducted in a designated laboratory experiment room. A redeveloped version of the system designed in a previous study was used [10–12]. Learners sat in front of a computer display and talked to each other orally. The experimental system was developed in the Java language and run on an in-house server-client network platform. The two learners' computers were connected through a local area network, and task execution was controlled by a program on the server. The system also features a conversational PCA that provided meta-cognitive suggestions [10] to facilitate their explanations. The example of the displays are shown in Fig. 1.

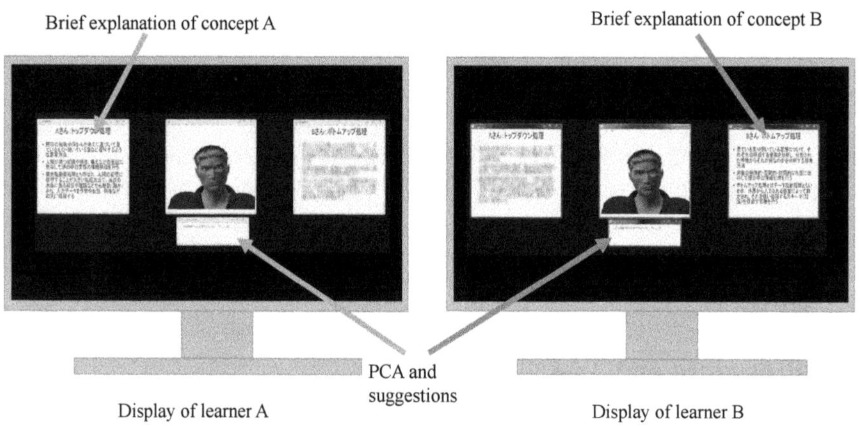

Fig. 1. Example of participants' screens.

An embodied PCA was presented in the center of the screen, which physically moved when it spoke. Below the PCA, there was a text box that showed messages. The experimenter sat to the side in the experiment room and manually instructed the PCA to provide meta-cognitive suggestions. The suggestions were made once per minute when there was a gap in conversation. Five types of meta-cognitive suggestions were used, such as reminding learners to achieve the task goal [2] and facilitating metacognition [11].

During the task, the experimenter also video-recorded the learner's facial expressions and recorded their voice during the main task. The facial recordings were used as one measure to understand the learner's affective state during their task, as explained further in the next section. All of the recorded conversations were transcribed into text to further analyze the quality of the explanations.

2.3 Measures

This study examined the performance of the participants on the pre- and post-tests and the quality of the collaboration through coding the learners' performance.

Pre-and Post-tests. The responses to the test were coded as follows: 0 = a wrong explanation, 1 = a naive but correct explanation, 2 = a concrete explanation based on the presented materials, and 3 = a concrete explanation based on the presented materials that used examples and metacognitive interpretations. Two coders conducted this analysis with an accuracy of 78%. They discussed any mismatching codes to determine the final codes.

Quality of Collaboration. The study adopted part of the coding scheme from [23] that are related to emotion capture. The original full scheme is as follows: 1 = mutual understanding, 2 = dialogue management, 3 = information pooling, 4 = reaching consensus, 5 = task division, 6 = time management, 7 = reciprocal interaction, and 8 = individual task orientation. The present study excluded the codes 3, 5, 6, and 7 because they are inappropriate for this study. Just as for the pre- and post-tests, two coders evaluated the transcripts of the experiment dialogues, with a coding match of 85%.

Facial Expressions. For the facial expression analysis, this study used Face Reader (https://www.noldus.com/) to evaluate the emotional states of the learners during the interactions. This system can classify expressions as one of the emotional categories of joy, anger, sadness, surprise, fear, or disgust [17]. The tool recognizes fine-grained facial features based on minimal muscular movements described by the Facial Action Coding System (FACS). The systems use Active Appearance Model (AAM) for creating a model of the facial expressions for classification, where the shape of the face is defined by a shape vector that contains the coordinates of the landmark points (for details of the algorithm, see [17]). In addition, the two coders checked the reliability of the automated coding by randomly selecting, manually coding, and checking the accuracy of the automatically detected emotional states. The accuracy of the recognition system was 72%. In this study, we calculated the variable of each emotional state for each individual and used it as a representative value for analysis. Using these values as predictors, we investigated how they can predict the learning performance and collaboration process.

3 Results

3.1 Performance on the Pre- and Post-tests

For each individual learner, the gain in score between the pre- and post-tests was calculated (gain = [post-test score] – [pre-test score]). Pearson's correlation coefficients were calculated to determine if there was correlation between the major detected states and the gain. Table 1 shows the results, which show that no correlation was detected between emotional state and performance gain.

Table 1. Results of correlation analysis for the performance and emotional states: "n.s." = no significance.

Neutral	Happy	Sad	Angry	Surprised	Scared	Disgusted
−0.183 n.s.	0.245 n.s.	−0.045 n.s.	0.077 n.s.	−0.138 n.s.	0.1 n.s.	−0.056 n.s.

3.2 Collaboration Process

To investigate the relationships among the learning process and emotional states, we conducted a Pearson's correlation analysis for the coded variables and each emotional state. Table 2 shows the results.

Table 2. Results of correlation analysis for the collaboration process and emotional states: "n.s." = no significance, "+" = marginal significance ($p < .10$), and "*" = significance ($p < .05$).

	Neutral	Happy	Sad	Angry	Surprised	Scared	Disgusted
1. Mutual understanding	−0.278 +	0.194 n.s.	−0.071 n.s.	0.302 *	0.056 n.s.	−0.226 +	0.088 n.s.
2. Dialogue management	−0.01 n.s.	−0.249 +	−0.177 n.s.	0.285 +	0.253 +	0.046 n.s.	−0.003 n.s.
4. Reaching consensus	−0.202 n.s.	0.340 *	−0.098 n.s.	0.301 *	−0.116 n.s.	−0.086 n.s.	0.021 n.s.
8. Individual task orientation	−0.4 *	0.348 *	−0.077 n.s.	0.283 n.s.	−0.052 n.s.	−0.065 n.s.	0.236 n.s.

The following sections further analyze the regressions for each type of collaborative process.

"Mutual Understanding". The Pearson's correlation analysis shows that there was a significant correlation between "mutual understanding" and "angry." As predicted, this could be due to the fact that learners working hard to develop mutual understandings in this jigsaw-like task, learners experienced more interpersonal conflict and expressed angry facial expressions. A multiple regression analysis was performed in which learning gain was regressed for each of the variables. The regression coefficient R^2 was .385 and the analysis of variance (ANOVA) F-value was 2.322, indicating significance for both variables (p = .05.) The results suggest that the degree to which the process of trying to establish common ground can be predicted by facial expressions displaying anger. This supports our hypothesis H1-a.

"Dialogue Management". The results of the Pearson's correlation analysis show that there were no significant correlations among any of the variables for "dialogue management." However, some marginal effects were found for the "happy," "angry," and "surprised" emotions.

"Reaching Consensus". The results of the Pearson's correlation analysis show that there were significant correlations between "reaching consensus" and "happy" and "angry" emotions. Learners may have experienced happiness because they had reached consensus. Anger may have appeared because learners experienced conflicts about their different perspectives and/or frustration prior to reaching consensus. To further investigate the prediction ability of these emotions, a multiple regression analysis was conducted. However, the regression coefficient R^2 was .316 and the ANOVA F-value was 1.717, indicating no significance for both variables (p = .14.)

"Individual Task Orientation". The results of the Pearson's correlation analysis show that there were significant correlations between "individual task orientation" and "happy" and "neutral." This indicates that when learners worked well individually they were engaging with the task with positivity. To further investigate the predictability, we conducted a multiple regression analysis where learning performance (the dependent variable) was regressed on each of the variables. The regression coefficient R^2 was .358 and the ANOVA F-value was 2.072, indicating only marginal significance (p = .08.)

4 Discussion and Conclusions

This study focused on learning situations in which learners were engaged in a jigsaw-based collaborative learning task, which requires interactive conflicts to achieve the goal. Using this task, the author investigated emotional states, which have been recently employed as important indicators for understanding learners' internal processes. Facial recognition technology was used to estimate the learners' facial expressions, which were then used to reveal the relationships between the social interactive process and learning performance while using a tutoring system. To investigate this, this study used a collaborative learning platform designed by the authors in which an embodied PCA was embedded. This study hypothesized that both positive and negative emotional states can capture the process of several types of interaction process such as developing common ground and furthermore learning performance. However, the results show that emotional states were useful to predict only for the learning process. More specifically, negative emotions detected from learners' facial expressions were able to predict the process of developing mutual understandings (supporting hypotheses H1-a.) This is considered that when developing consensus in this task, learners have to integrate their different perspectives and thus require interpersonal conflicts, which may be associated with confusions which can be detected as negative emotions.

Further investigation should be conducted by directly examining the degree of confusions such as dependent variables used by [6], for future work. Moreover, combinations of using several different variables should provide a larger view of the relationships between the interaction process and the types of emotional states of the learners.

On the other hand, none of the detected emotions were useful for predicting learning performance (not supporting hypotheses H1-b, and H2-b.)For this point, it can be discussed that gaining knowledge in this task was more-like an individual activity and there was hence no need for a learner to express his/her emotional state through facial expressions when thinking and reasoning. Some studies point that collaborative problem-solving is composed by phases of (1) task work which is to build internal knowledge and (2) team work (coordination) which is to exchange and share internalized knowledge to build collective knowledge [7]. It can be considered that each factors should be supported individually by using different types of facilitation methods [13]. Also, our hypotheses H1-b and H2-b might hold if we further investigate on other emotional or affective states using different measures.

In conclusion, these results will contribute to development of intelligent tutoring systems because the results show that learners' interaction process can be detected from emotional states. On designing such systems, detecting individuals' emotional states accurately is one of the challenges in artificial intelligence, but recently there has been many successes in the software development and this study is one good example. Past studies in ITS has not yet fully demonstrated how facial expressions can be automatically collected and used for detecting their emotional states especially in a task in a conflict-based collaborative learning environment. This paper contributes to ITS by showing empirical results based on laboratory study, showing how such technology enables to detect emotional states of learners conversational process. The next aim of this research is to develop a classifier based on our results and then develop a system that provides adaptive PCA facilitation based on the real-time recognition of the learner's states.

Acknowledgments. This work was supported by Grant-in-Aid for Scientific Research (C), Japan Society for the Promotion of Science (JSPS) KAKENHI Grant Number 16K00219.

References

1. Aronson, E., Patnoe, S.: The Jigsaw Classroom: Building Cooperation in the Classroom, 2nd edn. Addison Wesley Longman, New York (1997)
2. Azevedo, R., Cromley, J.: Does training on selfregulated learning facilitate students' learning with hypermedia? J. Educ. Psychol. **96**(3), 523–535 (2004)
3. Biswas, G., Leelawong, K., Schwartz, D., Vye, N.: Learning by teaching: a new paradigm for educational software. Appl. Artif. Intell. **19**(3), 363–392 (2005)
4. Chi, M., Leeuw, N., Chiu, M., Lavancher, C.: Eliciting self-explanations improves understanding. Cogn. Sci. **18**(3), 439–477 (1994)

5. von Davier, A.A., Hao, J., Liu, L., Kyllonen, P.: Interdisciplinary research agenda in support of assessment of collaborative problem solving: lessons learned from developing a collaborative science assessment prototype. Comput. Hum. Behav. **76**, 631–640 (2017)
6. D'Mello, S., Lehman, B., Pekrun, R., Graesser, A.: Confusion can be beneficial for learning. Learn. Instr. **29**, 153–170 (2014)
7. Fiore, S.M., Rosen, M.A., Smith-Jentsch, K.A., Salas, E., Letsky, M., Warner, N.: Toward an understanding of macrocognition in teams: predicting processes in complex collaborative contexts. Hum. Factors **52**, 203–224 (2010)
8. Graesser, A., Chipman, P., Haynes, B., Olney, A.: AutoTutor: an intelligent tutoring system with mixed-initiative dialogue. IEEE Trans. Educ. **48**(4), 612–618 (2005)
9. Graesser, A., McNamara, D.S.: Computational analyses of multilevel discourse comprehension. Top. Cogn. Sci. **3**(2), 371–398 (2011). https://doi.org/10.1111/j.1756-8765.2010.01081.x
10. Hayashi, Y.: On pedagogical effects of learner-support agents in collaborative interaction. In: Proceeding of the 11th International Conference on Intelligent Tutoring Systems (ITS 2012), pp. 22–32 (2012)
11. Hayashi, Y.: Togetherness: Multiple pedagogical conversational agents as companions in collaborative learning. In: Proceeding of the 12th International Conference on Intelligent Tutoring Systems (ITS 2014), pp. 114–123 (2014)
12. Hayashi, Y.: Coordinating knowledge integration with pedagogical agents: effects of agent gaze gestures and dyad synchronization. In: Proceeding of the 13th International Conference on Intelligent Tutoring Systems (ITS 2016), pp. 254–259 (2016)
13. Hayashi, Y.: Gaze feedback and pedagogical suggestions in collaborative learning. In: Nkambou, R., Azevedo, R., Vassileva, J. (eds.) ITS 2018. LNCS, vol. 10858, pp. 78–87. Springer, Cham (2018). https://doi.org/10.1007/978-3-319-91464-0_8
14. Hayashi, Y.: The power of a "maverick" in collaborative problem solving: an experimental investigation of individual perspective-taking within a group. Cogn. Sci. **42**(S1), 69–104 (2018). https://doi.org/10.1111/cogs.12587
15. Hesse, F., Care, E., Buder, J., Sassenberg, K., Griffin, P.: A framework for teachable collaborative problem solving skills. In: Griffin, P., Care, E. (eds.) Assessment and Teaching of 21st Century Skills. EAIA, pp. 37–56. Springer, Dordrecht (2015). https://doi.org/10.1007/978-94-017-9395-7_2
16. Koedinger, K.R., Anderson, J.R., Hadley, W.H., Mark, M.A.: Intelligent tutoring goes to school in the big city. Int. J. Artif. Intell. Educ. (IJAIED) **8**, 30–43 (1997)
17. Kuilenburg, v.H., Wiering, M., Uyl, d.M.: A model based method for automatic facial expression recognition. In: Proceedings of the European Conference on Machine Learning (ECML2005), pp. 194–205 (2005)
18. Kumar, R., Rose, C.: Architecture for building conversational architecture for building conversational agents that support collaborative learning. IEEE Trans. Learn. Technol. **4**(1), 21–34 (2011)
19. Lave, J., Wenger, E.: Situated Learning - Legitimate Peripheral Participation. Cambridge University Press, Cambridge (1991)
20. Leelawong, K., Biswas, G.: Designing learning by teaching agents: the Betty's brain system. Int. J. Artif. Intell. Educ. **18**(3), 181–208 (2008)
21. Matsuda, N., Yarzebinski, E., Keiser, V., Raizada, R., Stylianides, G.J., Koedinger, K.R.: Studying the effect of a competitive game show in a learning by teaching environment. Int. J. Artif. Intell. Educ. **23**(1), 1–21 (2013)

22. McNamara, D., O'Reilly, T., Rowe, M.: iSTART: A Web-Based Tutor that Teaches Self-explanation and Metacognitive Reading Strategies. Lawrence Erlbaum Associates, Mahwah (2007)
23. Meier, A., Spada, H., Rummel, N.: A rating scheme for assessing the quality of computer-supported collaboration processes. Int. J. Comput.-Support. Collab. Learn. **2**(1), 63–86 (2007). https://doi.org/10.1007/s11412-006-9005-x
24. Metcalfe, J., Wiebe, D.: Intuition in insight and noninsight problem solving. Mem. Cogn. **15**(3), 238–246 (1987)
25. Miyake, N.: Constructive interaction and the interactive process of understanding. Cogn. Sci. **10**(2), 151–177 (1986)
26. OECD: PISA 2015 Results (Volume V): Collaborative Problem Solving. OECD Publishing, Paris (2017). https://doi.org/10.1787/9789264285521-en
27. Shirouzu, H., Miyake, N., Masukawa, H.: Cognitively active externalization for situated reflection. Cogn. Sci. **26**(4), 469–501 (2002)
28. VanLehn, K., Graesser, A.C., Jackson, G.T., Jordan, P., Olney, A., Rosé, C.P.: When are tutorial dialogues more effective than reading? Cogn. Sci. **31**(1), 3–62 (2007). https://doi.org/10.1080/03640210709336984
29. Vygotsky, L.S.: The Development of Higher Psychological Processes. Harvard University Press, Cambridge (1980)

Fact Checking Misinformation Using Recommendations from Emotional Pedagogical Agents

Ricky J. Sethi(✉), Raghuram Rangaraju, and Bryce Shurts

Fitchburg State University, Fitchburg, MA 01420, USA
rickys@sethi.org

Abstract. Dealing with complex and controversial topics like the spread of misinformation is a salient aspect of our lives. In this paper, we present initial work towards developing a recommendation system that uses crowd-sourced social argumentation with pedagogical agents to help combat misinformation. We model users' emotional associations on such topics and inform the pedagogical agents using a recommendation system based on both the users' emotional profiles and the semantic content from the argumentation graph. This approach can be utilized in either formal or informal learning settings, using threaded discussions or social networking virtual communities.

1 Introduction

Dealing with *complex* and *controversial* questions like "Do we use only 10% of our brain?" or "Were the crowds truly larger at Donald Trump's inauguration than at Barack Obama's?" has always been a salient aspect of our lives. Ironically, such *emotionally* charged topics often elicit the **backfire effect**, where people's opinions harden in the face of facts to the contrary [8].

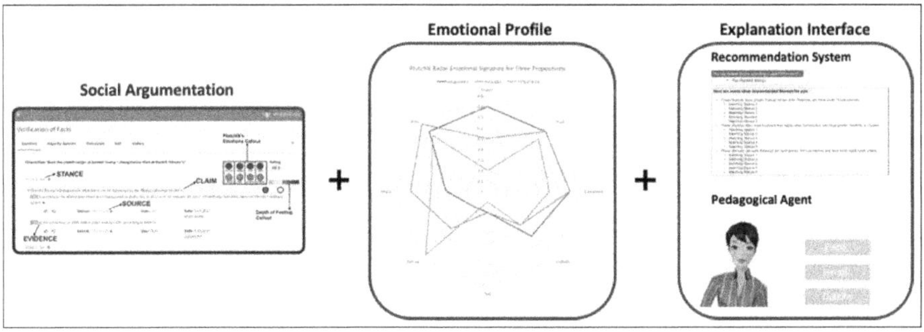

Fig. 1. System Overview: Social Collaborative Argumentation + Emotional Profiles + [Multi-Attribute Utility Theory Recommendation System + Pedagogical Agents]

A. Coy et al. (Eds.): ITS 2019, LNCS 11528, pp. 99–104, 2019.
https://doi.org/10.1007/978-3-030-22244-4_13

In this paper, we present initial work towards developing a recommendation system that uses crowd-sourced social argumentation with pedagogical agents (PAs) to help engender an analytical, evidence-based approach for combating misinformation. Our goal is to help learners *think critically* about such topics by using social collaborative argumentation supported by PAs that are informed by a recommendation system based on user' emotional and semantic profiles. An overview of our approach is shown in Fig. 1.

1.1 Background

Our goal is to help people more effectively explore and understand their possibly subconscious biases in an effort to overcome the backfire effect and formulate more varied insights into complex topics. We utilize a *structured learning environment*, consisting of a **virtual community** that supports **social collaborative argumentation**, to build an argumentation graph which captures the semantic content of the propositions associated with such topics [9,10].

We then address the role of emotion in reasoning by modeling the **emotional profiles** of users and contributors. We use both self-reported emotions and natural language processing with sentiment analysis from an **explanation interface** to create emotional profiles for users [8].

We now extend this approach to incorporate **Pedagogical Agents** (PA) to aid users' learning and gauge the impact of different PAs exhibiting varying degrees of emotion and knowledge. The PAs are informed by a **recommendation system** that fuses both the emotional profiles of users and the semantic content from the argumentation graph. We can use this emotional assessment to also gauge the extent of the backfire effect and the change in critical thinking as well as the ability of users to monitor and regulate their self-regulated learning processes by considering different types of information, evidence, and social influence delivered by the PAs.

This approach can have broad applicability for improving *online classroom learning* that utilizes threaded discussions; for *facilitating decision-making* amongst domain experts; and for *creating an informed electorate* that can assess the trustworthiness of information and information sources and lessen the risks of untrustworthy information like "alternative facts" and "fake news." It can also be used to *alter existing social networking sites* like Facebook to leverage their current userbase and aid in mitigating destructive online behaviour like spreading misinformation.

The three parts of our overall approach, shown in Fig. 1, consist of the:

1. Social Collaborative Argumentation and Virtual Community
   Our social collaborative argumentation framework allows arguments to be expressed in a graph structure with input to be provided by a crowd that is mediated by experts. The fundamental idea is to incorporate content, ratings, authority, trust, and other properties in a graph-based framework combined with a recommendation system which explains the tradeoff of various competing claims based on those attributes [9–12].

2. Emotional and Semantic Profiles
   We model users' emotional associations on complex, controversial topics as well as create a proposition profile, based on the semantic and collaborative content of propositions. Our framework combines emotional profiles of users for each proposition along with the semantic proposition profile [8,9].
3. Explanation Interface Recommendation System and Pedagogical Agents Interaction, as developed next.

2 Recommender System Explaining Interface

Long, complex arguments have a tendency to overwhelm users and leave them unable to decide upon which Stance is best supported for complex or controversial topics [2]. Recommendation systems can help target content to aid users' decision making and come in many varieties [3,14].

Although there are advantages to these models and other explanation interfaces [14], including case-based and knowledge-based, combating misinformation requires the ability to not just look deeply but to look laterally and be able to account for multiple attributes [15]. Multi-Attribute Utility Theory (MAUT) [7] based explanation systems have been shown to do exactly this by increasing trustworthiness and persuasiveness in users' decision-making [4,5].

Since our goal is to also increase the trustworthiness and persuasiveness of examining information online, we use the MAUT-based approach to create a novel explanation interface that can help users reason about controversial or nuanced topics comprehensively, including analyzing the semantic and emotional content of a complex argument in order to overcome both cognitive and emotional biases that contribute to echo chambers and the backfire effect. As such, not only does the system need to address the trust and authority of the information and its sources but the Viewer needs to be able to assess these components independently, as well [9,10].

We therefore create a recommender system that can recommend different Stances for a Proposition depending on the emotional and cognitive content of the collaborative argument. We anticipate that most Propositions will represent complex, nuanced Topics and so will have a number of Stances in general. The system will thus suggest supported Stances based on weights from the Viewer.

We do so by forming both Semantic and Emotional Profiles. Analogous to the standard recommendation models, we map {Products} $\rightarrow$ {Stances} and {Attributes} $\rightarrow$ {Claims, Evidence, Sources, Contributors}. Wherein the traditional recommender system utilizes the content of {Reviews}, we consider the content of {Claims, Evidence} nodes in the argumentation graph, G_A.

We also conduct a sentiment analysis on these Claims and Evidence nodes to create a Proposition Sentiment Profile. Using Viewer ratings of the emotional and depth of feeling callouts, we construct a Viewer Emotional Profile as well as a Proposition Emotional Profile [8,9]. We further construct a Proposition Semantic Profile by using a Text Analysis approach for the semantic content and combining it with the collaborative cognitive content as well as the weights along the various dimensions of Contributor ratings, trust, and authority encapsulated in the edges $e \in E$ of the G_A. Finally, we create the following utility model as per MAUT [7]:

$$U_v(S) = \sum_{i=1}^{m} w_i \cdot [\alpha \cdot V_i(S) + (1 - \alpha) \cdot O_i(S)] + \sum_{j=m+1}^{n} w_j \cdot O_j(S) \quad (1)$$

where $U_v(S)$ is the utility of a Stance, S, for a Viewer, v, in terms of the Viewer's preferences. This contains an attribute model value, V_i, which denotes the Viewer's preference for each attribute, a_i, as well as a sentiment model value based on opinion mining, or sentiment analysis, O, as detailed next. The Sentiment Stance Model is expressed as:

$$O_i(S) = \frac{1}{|R(a_i, S)|} \sum_{r \in R(a_i, S)} O_i(r) \quad (2)$$

where $R(a_i, S)$ is the set of Claims and Evidence, analogous to Reviews in traditional recommender systems, with respect to a Stance, S, that contain the opinions extracted on attributes, a_i, which consist of Claims, Evidence, Sources, and Contributors. The sentiment for each review component, $O_i(r)$, is defined as:

$$O_i(r) = \frac{\sum_{e \in E(a_i, r)} polarity(e)^2}{\sum_{e \in E(a_i, r)} polarity(e)} \quad (3)$$

where E is the set of sentiment elements associated with all the features mapped to an attribute a_i in a review component r and $polarity(e)$ is the polarity value for the element e as derived using standard statistical sentiment analysis [6].

The utility model, $U_v(S)$, establishes a Viewer's preferences using a standard weighted additive form of the value functions [4,5,7]. This model allows us to calculate the tradeoffs in order to explicitly resolve a Viewer's preferences. We calculate this tradeoff among the top k recommended Stance candidates; these are limited to $min[5, |S|]$, where 5 is the optimum number of options in a category less than 6 as discovered by [4], and $|S|$ is the number of returned Stances. We then run the Apriori algorithm [1] over all candidates' tradeoff vectors in order to calculate the Categories for Stances, as motivated by [4,5].

All Stances with the same subset of tradeoff pairs are grouped together into one Category; these tradeoff vectors take the form of sentiment and feature combined into a phrase like, "more Evidence", "more Polarized", etc. The Categories, in turn, use the tradeoff vectors as their descriptors so that we end up with Category Titles like, "This Stance has more Evidence from highly-rated Contributors and have greater similarity in terms of Content." The category title is thus the explanation that shows the advantages and disadvantages of the Stances. A mockup of how this would appear is shown in Fig. 2.

Finally, we want category titles that are different to maximize how informative they are. As such, we also map the Diversity, D, of each Category, c, in the

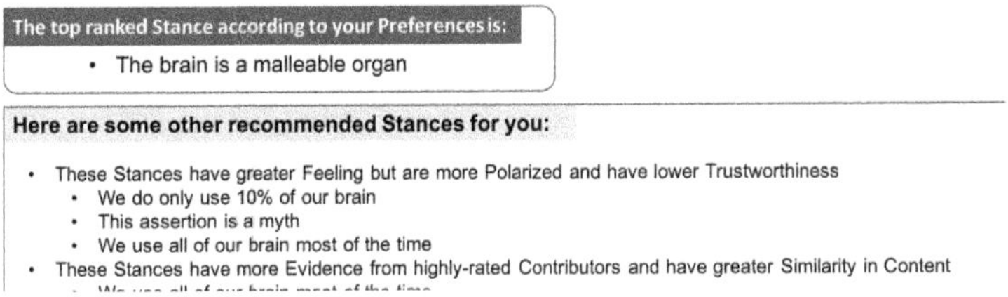

Fig. 2. A sample of the recommendations for Stances organized according to Category Titles made up of tradeoff vectors. Tradeoff vectors are of the type, "more Evidence", "lower Trustworthiness", etc., while Category Titles are of the form, "These Stances have greater Feeling but are more Polarized and have lower Trustworthiness".

set of Categories, C, in terms of both the Category Title, which is simply the set of tradeoff vectors, and the set of Stances in c, $S(c)$, as:

$$D(c, S(c)) = \min_{c_i \in C}[(1 - \frac{c \cap c_i}{|c|}) \times (1 - \frac{S(c) \cap S(c_i)}{|S(c)|})] \quad (4)$$

In the last step, the Viewer can update their preferences by using a button to offer better matching Stances. By also incorporating our structured discussion metrics [13], this framework can be applied to everything from analyzing misinformation to structuring discussions in online courses to ensure the information is trustworthy and the information sources assessable via the Category Titles.

3 Pedagogical Agents (PAs)

This explanatory recommendation infrastructure also utilizes PAs which will guide the user's experience. For example, suppose the topic a user wants to analyze is the crowd size at the 2017 inauguration. Viewers can examine the social argument with the aid of a PA. They can choose the kind of PA with whom to interact, like Friendly Republican or Objective Democrat or Angry Independent, etc.

In our approach, an argument is composed of Stances, Claims, Evidence, and Sources. This crowd-sourced argument is built out by the contributors to our system [8,9]. Once the argument is constructed, Viewers can interact with the argument and the PA. This PA can then guide the Viewer through the examination of the argument using the biases captured in the PA and help critically analyze the topic. Viewers can change the PAs in their preferences or they can change them depending on affective data, either implicit or explicit, as collected by our framework.

In particular, the PAs can help viewers navigate the claims and especially those that might contradict their initial stance on a topic by helping them evaluate both the evidence and feelings for alternative claims. Since the PA is built upon the Multi Attribute Utility Theory, it will be able to provide the explanation for why the alternative claims are presented from both a cognitive and affective perspective.

Acknowledgments. We would like to gratefully acknowledge support from the Amazon AWS Research Grant program. We would also like to thank Roger Azevedo for the valuable discussions and support.

References

1. Agrawal, R., Imieliński, T., Swami, A.: Mining association rules between sets of items in large databases. In: Acm Sigmod Record, vol. 22, pp. 207–216. ACM (1993)
2. Baram-Tsabari, A., Sethi, R.J., Bry, L., Yarden, A.: Asking scientists: a decade of questions analyzed by age, gender, and country. Sci. Educ. **93**(1), 131–160 (2008). https://doi.org/10.1002/sce.20284
3. Chen, L., Chen, G., Wang, F.: Recommender systems based on user reviews: the state of the art. User Model. User-Adapt. Interact. **25**(2), 99–154 (2015)
4. Chen, L., Pu, P.: Experiments on the preference-based organization interface in recommender systems. ACM Trans. Comput.-Hum. Interact. **17**(1), 1–33 (2010). https://doi.org/10.1145/1721831.1721836
5. Chen, L., Wang, F.: Sentiment-enhanced explanation of product recommendations. In: Proceedings of the 23rd International Conference on World Wide Web, pp. 239–240. ACM (2014)
6. Hu, M., Liu, B.: Mining and summarizing customer reviews. In: Proceedings of the Tenth ACM SIGKDD International Conference on Knowledge Discovery and Data Mining, pp. 168–177. ACM (2004)
7. Keeney, R.L., Raiffa, H., Rajala, D.W.: Decisions with multiple objectives: preferences and value trade-offs. IEEE Trans. Syst. Man Cybern. **9**(7), 403–403 (1979)
8. Sethi, R., Rangaraju, R.: Extinguishing the backfire effect: using emotions in online social collaborative argumentation for fact checking. In: 2018 IEEE International Conference on Web Services, ICWS 2018, San Francisco, CA, USA, 2–7 July 2018, pp. 363–366 (2018). https://doi.org/10.1109/ICWS.2018.00062
9. Sethi, R.J.: Crowdsourcing the verification of fake news and alternative facts. In: ACM Conference on Hypertext and Social Media (ACM HT) (2017). https://doi.org/10.1145/3078714.3078746
10. Sethi, R.J.: Spotting fake news: a social argumentation framework for scrutinizing alternative facts. In: IEEE International Conference on Web Services (IEEE ICWS) (2017)
11. Sethi, R.J., Bry, L.: The Madsci network: direct communication of science from scientist to layperson. In: International Conference on Computers in Education (ICCE) (2013)
12. Sethi, R.J., Gil, Y.: A social collaboration argumentation system for generating multi-faceted answers in question & answer communities. In: CMNA at AAAI Conference on Artificial Intelligence (AAAI) (2011)
13. Sethi, R.J., Rossi, L.A., Gil, Y.: Measures of threaded discussion properties. In: Intelligent Support for Learning in Groups at International Conference on Intelligent Tutoring Systems (ITS) (2012)
14. Tintarev, N., Masthoff, J.: Explaining recommendations: design and evaluation. In: Ricci, F., Rokach, L., Shapira, B. (eds.) Recommender Systems Handbook, pp. 353–382. Springer, Boston, MA (2015). https://doi.org/10.1007/978-1-4899-7637-6_10
15. Wineburg, S., McGrew, S.: Lateral reading: reading less and learning more when evaluating digital information (2017)

Intelligent On-line Exam Management and Evaluation System

Tsegaye Misikir Tashu[1,2(✉)], Julius P. Esclamado[4], and Tomas Horvath[1,3]

[1] Faculty of Informatics, Department of Data Science and Engineering, Telekom Innovation Laboratories, ELTE-Eötvös Loránd University, Pázmány Péter Sétány 1117, Budapest, Hungary
{misikir, tomas.horvath}@inf.elte.hu

[2] Faculty of Informatics, 3in Research Group, ELTE-Eötvös Loránd University, Martonvásár, Hungary

[3] Faculty of Science, Institute of Computer Science, Pavol Jozef Šafárik University, Jesenná 5, 040 01 Košice, Slovakia

[4] Cagayan de Oro, Philippines
esclamadojp@gmail.com

Abstract. Educational assessment plays a central role in the teaching-learning process as a tool for evaluating students' knowledge of the concepts associated with the learning objectives. The evaluation and scoring of essay answers is a process, besides being costly in terms of time spent by teachers, what may lead to inequities due to the difficulty in applying the same evaluation criteria to all answers. In this work, we present a system for online essay exam evaluation and scoring which is composed of different modules and helps teachers in creating, evaluating and giving textual feedbacks on essay exam solutions provided by students. The system automatically scores essays, semantically, using pair-wise approach. Using the system, the teacher can also give an unlimited number of textual feedbacks by selecting a phrase, a sentence or a paragraph on a given student's essay. We performed a survey to assess the usability of the system with regard to the time saved during grading, an overall level of satisfaction, fairness in grading and simplification of essay evaluation. Around 80% of the users responded that the system helps them to grade essays more fairly and easily.

Keywords: Automatic essay evaluation · Automatic feedback · Intelligent tutoring · Learning assessment

1 Introduction

The most common and traditional way of evaluating and scoring exams is done using the pen and pencil system. Paper-based examinations and pen-and-pencil evaluation are prone to different kinds of problems such as inconsistency among markers because of fatigue, loss of concentration arising from boredom and neatness of student handwriting; variables in the markers' background [1]; and the time it takes to complete the

J. P. Esclamado—Independent Researcher

A. Coy et al. (Eds.): ITS 2019, LNCS 11528, pp. 105–111, 2019.
https://doi.org/10.1007/978-3-030-22244-4_14

marking process. These factors are especially present in case of evaluating solutions subjective types of exams where the evaluation is not based on some objective criteria but on teacher's subjective opinion. Various on-line exam evaluation systems have been developed, however, these are mostly focusing on objective types of exams and achieve good performance [2, 3]. For subjective types of exams, Automatic Essay Evaluation (AEE) systems have been introduced [4, 5]. The main focus of this paper is on e-Testing systems with particular focus on AEE.

After investigating the current AEE systems and the state-of-the-art in e-Testing systems, we have designed and implemented an enhanced web-based e-Testing system that helps both the teachers and students in the process of examination and scoring[1]. It allows the assessors to give feedback in the form of textual comments on selected parts of student's submissions and scores the exams automatically. The features and functionalities of the system is explained in Sect. 3. The rest of this paper is organized as follows: Sect. 2 provides an overview of the existing works and approaches and Sect. 3 provides the details of the system. AEE method is presented in Sect. 4. Section 5 presents evaluation results and Sect. 6 presents concluding remarks.

2 Related Works

AEE is an interesting development domain that has been ongoing since the 1960s up to today [6]. AEE systems are distinguished from each other primarily the way they evaluate essays such that either by style or by content or both. Another distinction criterion is the approach adopted for assessing style and/or content. The most important approaches found in the literature of AEE systems are Statistical, Latent Semantic Analysis (LSA), Natural Language Processing (NLP), Machine Learning (ML) and Artificial Neural Network (ANN). AEE systems that focused on statistical approaches capture only the structural similarity of texts. The following systems, based on LSA, did more than a simple analysis of co-occurring terms. They introduced two new approaches to the problem: a comparison based on a corpus and an algebraic technique that allowed identifying similarities between two texts with different words [7]. The latest systems are based on NLP, ML and ANN techniques and can do intelligent analyzes that capture the semantic meaning of an essay. As mentioned above, the distinction is made between grading essays based on content and those based on style. AEE systems that evaluate and score primarily based on style is the Project Essay Grade (PEG) [6] using linguistic features via proxies. Systems utilizing content are Intelligent Essay Assessor (IEA) [8] using LSA, Educational Testing Service [5] using NLP, and, Pair-wise [4] based on ANN.

Using the current AEE systems, evaluating and scoring essay can be done automatically but the assessor cannot interact with and give feedback to student answers in the form of textual comments. The students also do not have the possibility to learn from their mistakes as AEE systems do not have such a feature so far. To support both scoring and providing feedback in the form of comment, we designed and implemented web-based intelligent exam management and evaluation system.

---

[1] http://www.etestsupport.com.

3 System Modules and Implementation

The System Management Module (Maintenance) includes the User information, Audit logs, User management and System security sub-modules. Using User Management, the system administrators can create and set the basic information for teachers and students. Using User management and Authentication, the administrator can set the identities of the teachers and students, maintain user information, grant access rights to the different modules for each user, backup and restore the system information. The Audit logs module is used to trace the activities and interaction with the system of each user. If something went wrong, the audit logs will be used to restore the system into a normal state. Figure 1 shows the available main and sub-modules of the system.

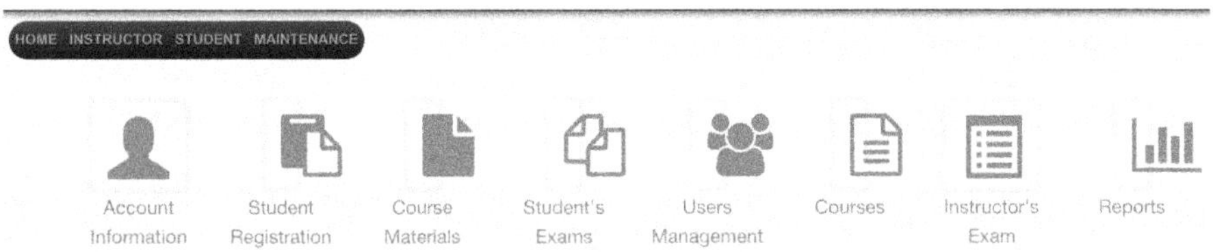

Fig. 1. The main and sub-modules of the e-Test system.

The Instructors module allows teachers to maintain their profile, change their log-in credentials, to add and maintain their own courses, to approve the requests from students to register for the course(s) and upload course materials[2]. The Student module allows students to register, obtain and maintain their log-in credentials, search and send a request for course registration to the teacher, submit solutions to exams and view their obtained score and feedback. The Exam module allows teachers to create private and public, subjective and objective types of exam. Teachers can also add students into their exam, dispense the exam online and set the starting and ending time of the exam. Once the exams are released by the teacher, the students can log-in to the system and can submit their solution within the given time-stamp. The students will receive a machine score for essay exams automatically.

The Feedback module is designed to be used by the teachers to give textual feedback to essay exam solutions. Using this module, the teacher can open each student's essay, select and highlight phrases, sentences or paragraphs and write textual comments to these highlighted parts, as depicted in Fig. 2. It allows the teachers to give an unlimited number of textual comments on a single essay solution. They can also modify the machine score based on their reviews. The students can also see the textual feedbacks given by their teacher.

[2] The developed system is not intended to serve as a learning management system (e.g. Moodle) but, rather, as an e-Testing framework. However, we have implemented this ability to upload course materials for teachers not using any other system.

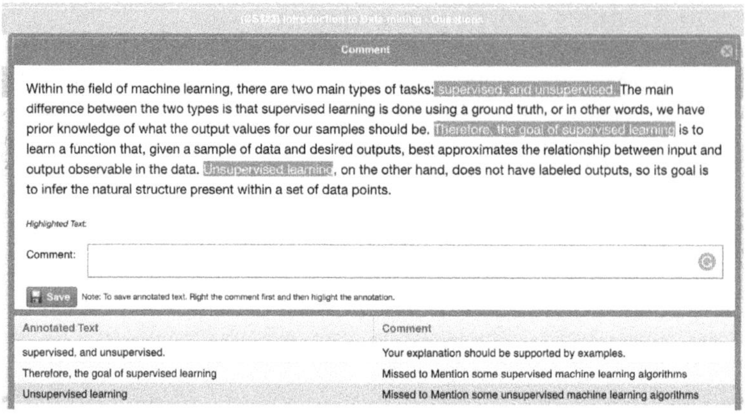

Fig. 2. Feedback module: showing how the teachers can give textual comments

The system was developed using web application development tools like Extended JavaScript (EXTJS), Hypertext Markup Language (HTML5.0), Cascading Style Sheet (CSS), Hypertext Pre-processor (PHP) and MySQL.

4 Automatic Scoring Process

The automatic essay evaluation and scoring process of the system is based on computing semantic similarity. For every essay question, the teacher should provide a reference answer (RA). The RA and the student answer (SA) are provided to the AEE algorithm as an input. The two inputs will be passed through the text pre-processing phase and the following activities will be carried out: tokenization; text-normalization; stopword removal and word lemmatization. The similarity between the RA and SA is computed using pair-wise method [4] which uses a neural word embedding to semantically represent the inputs and the score is computed using the Word Mover's distance (WMD) algorithm [9] redefined using cosine similarity. Given a RA and SA the cosine similarity between RA and SA is defined as follows

$$cosine\_sim(RA, SA) = \frac{Xi \times Xj}{\|Xi\| \times \|Xj\|} \tag{1}$$

where Xi is a vector representation of RA and Xj is a vector representation of SA.

The WMD utilizes the property of word2vec embeddings [10]. Therefore, the similarity between RA and SA is the maximum cumulative similarity that word vectors from document RA travels to match exactly to word vectors of document SA. In this regard, in order to compute the semantic similarity between SA and RA, SA will be mapped to RA using a pre-trained word embedding model[3]. Let **SA** and **RA** be nBOW

[3] https://code.google.com/archive/p/word2vec/.

representations of SA and RA, respectively. Let $T \in R^{nxn}$ be a flow matrix, where $T_{ij} \geq 0$ denotes how much the word w_i in SA has to "travel" to the word w_j in RA, and n is the number of unique words appearing in SA and RA. To transform **SA** to **RA** entirely, we ensure that the complete flow from the word w_i equals d_i and the incoming flow to the word w_j equals d'_j. The WMD is defined as follows using cosine similarity measure:

$$\max_{T \geq 0} \sum\nolimits_{i,j=1}^{n} Tij\, cosine\_sim(wi, wj) \tag{2}$$

$$\text{Subject to} \qquad \sum\nolimits_{j=1}^{n} Tij = di, \forall i \in \{1, \ldots, n\}$$

5 System Evaluation

We asked a series of questions to 20 instructors teaching at Wollo University, Faculty of Informatics, who used the system in their teaching process to evaluate the system but the survey did not include the feedback from the students' side yet. The results of the survey are reported in figure Fig. 3 and Table 1. In all of the figures, we ignored zero-response answer choices in the graph. The majority of the users replied well and, particularly, they expressed that the system meets their needs. The majority of the users also said that the system is useful and reliable. 60% of the respondents report that they are satisfied with the features of the system and meets their needs. The details for the five survey are reported in the graphs and tables below alongside the survey questions.

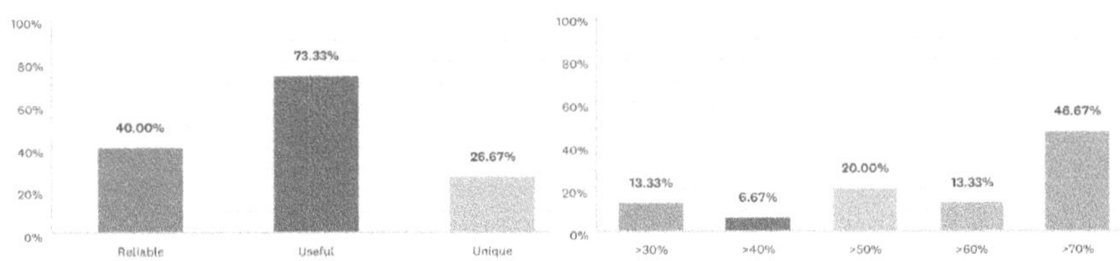

Fig. 3. Survey results collected from two questions (Left-which of the following words would you use to describe the system? Select all that apply. Right-how much time do you save while evaluating essay questions using the system?)

The performance of the pair-wise AEE was evaluated and compared to other state-of-the-art methods using root mean scored error. The dataset provided by Kaggle[4] is used to see the performance of the method at large scale. The results show that the pair-wise approach achieved lower level of error 13.31% while 15.35% and 19.23% respectively for latent semantic analysis and wordnet.

---

[4] https://www.kaggle.com/c/asap-sas.

Table 1. Survey results collected from the three questions which have the same answer choices (Q1-does the system help you grade more fairly? Q2-does the system save you time in grading? and Q3-does the system make the evaluation of essay questions more easily?)

Answer choices	Q1 responses	Q2 responses	Q3 responses
Strongly agree	47.67%	33.33%	33.33%
Agree	33.33%	45.00%	40.00%
Neutral	12.33%	15.00%	20.00%
Disagree	6.67%	6.67%	0.00%

6 Conclusions

A web-based online essay exam management and evaluation system is designed and implemented as a tool the teachers in creating and evaluating essay exams. The system allows the teacher to give textual feedbacks in the form of comment by selecting parts of the essay solution which he/she assumes is not correct. The system has five main modules and these are: System Management module, Instructors module, Students module, Exam and Feedback module. The automatic evaluation and scoring of essay answers are performed using pair-wise AEE method which uses a RA and SA to compute the semantic similarity and compute the score according to the weight of the question. In a survey, conducted to assess the feasibility and usability of the system, teachers responded that the system helps them to provide higher quality feedback in a short time. The preliminary results are promising and show good indications for further improvement of the developed system. The system at this level can only be used for English language essay exam and we did not perform the survey to measure how the system meets the needs and requirements for the students.

References

1. Leckie, G., Baird, J.-A.: Rater effects on essay scoring: a multilevel analysis of severity drift, central tendency, and rater experience. J. Educ. Meas. **48**, 399–418 (2011)
2. Kaya, B.Y., Kaya, G., Dagdeviren, M.: A sample application of web based examination system for distance and formal education. Procedia - Soc. Behav. Sci. **141**, 1357–1362 (2014)
3. Yağci, M., Ünal, M.: Designing and implementing an adaptive online examination system. Procedia - Soc. Behav. Sci. **116**, 3079–3083 (2014)
4. Tashu, T.M., Horvath, T.: Pair-wise: automatic essay evaluation using word mover's distance. In: The 10th International Conference on Computer Supported Education - Volume 2: CSEDU, pp. 59–66. SciTePress (2018)
5. Attali, Y.: A differential word use measure for content analysis in automated essay scoring. ETS Res. Rep. Ser. **36**, 1–19 (2011)
6. Page, E.B.: Grading essays by computer: progress report. In: Invitational Conference on Testing Problems (1966)

7. Thomas, P., Haley, D., deRoeck, A., Petre, M.: e-Assessment using latent semantic analysis in the computer science domain: a pilot study. In: Proceedings of the Workshop on eLearning for Computational Linguistics and Computational Linguistics for eLearning, pp. 38–44. ACL (2004)
8. Foltz, P.W., Laham, D., Landauer, T.K.: Automated essay scoring : applications to educational technology. In: World Conference on Educational Multimedia, Hypermedia and Telecommunications (ED-MEDIA) (1999)
9. Kusner, M.J., Sun, Y., Kolkin, N.I., Weinberger, K.Q.: From word embeddings to document distances. In: International Conference on Machine Learning (2015)
10. Mikolov, T., Chen, K., Corrado, G., Dean, J.: Distributed representations of words and phrases and their compositions. In: NIPS (2013)

Learning by Arguing in Argument-Based Machine Learning Framework

Matej Guid(✉), Martin Možina, Matevž Pavlič, and Klemen Turšič

Faculty of Computer and Information Science, University of Ljubljana, Ljubljana, Slovenia
matej.guid@fri.uni-lj.si

Abstract. We propose an approach for the development of argument-based intelligent tutoring systems in which a domain that can be successfully addressed by supervised machine learning is taught in an interactive learning environment. The system is able to automatically select relevant examples and counter-examples to be explained by the students. The students learn by explaining specific examples, and the system provides automated feedback on students' arguments, including generating hints. The role of an argument-based intelligent tutoring system is then to train the students to find the most relevant arguments. The students learn about the high-level domain concepts and then use them to argue about automatically selected examples. We demonstrate our approach in an online application that allows students to learn through arguments with the goal of improving their understanding of financial statements.

Keywords: Intelligent tutoring systems · Learning by arguing · Argument-based machine learning · Automated feedback generation · Financial statements

1 Introduction

When students collaborate in argumentation in the classroom, they are arguing to learn. Argumentation can help learners to accomplish a wide variety of important learning goals. [2] It involves elaboration, reasoning, and reflection. These activities have been shown to contribute to deeper conceptual learning [3]. Effective learning by arguing involves making knowledge explicit: learners that provide explanations, or make explicit the reasoning underlying their problem solving behavior, show the most learning benefits [4].

It is often the case that domains of interest can be represented by numerous learning examples. These learning examples can typically be described by various features and may contain labels such as 'good', 'bad' or 'ugly'. A domain described by labeled learning examples can be tackled by *supervised machine learning* algorithms. Supervised machine learning technique in which the algorithm attempts to label each learning example by choosing between two or more different classes is called *classification*. For instance, a classification task would

A. Coy et al. (Eds.): ITS 2019, LNCS 11528, pp. 112–122, 2019.
https://doi.org/10.1007/978-3-030-22244-4_15

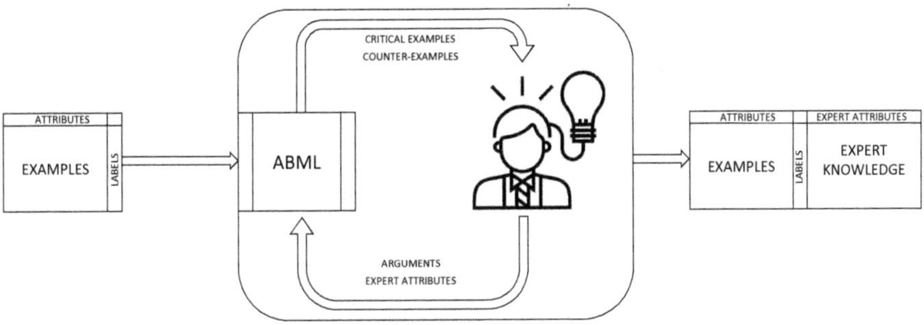

Fig. 1. ABML knowledge refinement loop with the domain expert.

be to learn a prediction model that distinguishes between successful and less successful companies based on their financial statements.

We propose an approach to the development of argument-based intelligent tutoring systems using argument-based machine learning (ABML) [12] framework. In principle, it allows any domain that can be successfully addressed by supervised machine learning to be taught in an interactive learning environment. It has been shown that ABML provides the opportunity to develop interactive teaching tools that are able to automatically select relevant examples and counter-examples to be explained by the student [15]. We extend that approach with automated feedback on students' arguments, including the generation of hints. In addition, we have developed a web application that can be easily adapted to various learning domains. It allows students to learn by arguing with a very clear objective: to improve their understanding of particular learning domains.

The chosen experimental domain was financial statement analysis. More concretely, estimating credit scores or the creditworthiness of companies. Our aim was to obtain a successful classification model for predicting the credit scores and to enable the students to learn about this rather difficult subject.

In the experiments, both the teacher and the students were involved in the interactive process of knowledge elicitation based on the ABML paradigm, receiving the feedback on their arguments. The aim of the learning session with the teacher was in particular to obtain advanced concepts (features) that describe the domain well, are suitable for teaching and also enable successful predictions. This was done with the help of a financial expert. In the tutoring sessions, the students learned about the intricacies of the domain and looked for the best possible explanations for automatically selected examples, using the teacher's advanced concepts in their arguments. The system is also able to track the student's progress in relation to these selected concepts.

2 Argument-Based Machine Learning

Argument-based machine learning (ABML) [12] is machine learning, extended by concepts from argumentation. In ABML, arguments are typically used as a means for users (e.g. domain experts, students) to elicit some of their knowledge by explaining the learning examples. The users only need to concentrate on one specific case at a time and impart knowledge that seems relevant for this case. They provide the knowledge in the form of arguments for the learning examples and not in the form of general domain knowledge.

ABML Knowledge Refinement Loop [11] enables an interaction between a machine learning algorithm and a domain expert (see Fig. 1). It is a powerful knowledge acquisition tool capable of acquiring expert knowledge in difficult domains [7–9,13]. The loop allows the expert to focus on the most critical parts of the current knowledge base and helps him to discuss automatically selected relevant examples. The expert only needs to explain a single example at the time, which facilitates the articulation of arguments. It also helps the expert to improve the explanations through appropriate counter-examples.

We use the ABCN2 [12] method, an argument-based extension of the well-known CN2 method [5], which learns a set of unordered probabilistic rules from examples with attached arguments, also called *argumented examples.*

2.1 ABML Knowledge Refinement Loop

In this section, we give a brief overview of the steps in the ABML knowledge refinement loop from the perspective of the expert:

Step 1: Learn a hypothesis with ABCN2 using the given data.
Step 2: Find the "most critical" example and present it to the student. If a critical example cannot be found, stop the procedure.
Step 3: Expert explains the example; the explanation is encoded in arguments and attached to the critical example.
Step 4: Return to step 1.

In the sequel, we explain (1) how to select critical examples and (2) how to obtain all necessary information for the selected example.

Identifying Critical Examples. The arguments given to the critical examples cause ABCN2 to learn new rules that cover these examples. A critical example is an example with a high probabilistic prediction error. The probabilistic error can be measured in different ways. We use the Brier Score with a k-fold cross-validation repeated n times (e.g. $n = 4, k = 10$), so that each example is tested n times. The most problematic example is therefore the one with the highest average probabilistic error over several repetitions of the cross-validation procedure.

Improving a Expert's Arguments. In the third step of the above algorithm, the expert is asked to explain the critical example. With the help of the expert's arguments, ABML will sometimes be able to explain the critical example, while sometimes this is still not entirely possible. Then we need additional information from the expert where the counter-examples come into play. The following five steps describe this idea:

Step 3a: Explain the critical example. The expert is asked the following question: "Why is this example in the class as given?" The answer can be either "I don't know" (the expert cannot explain the example) or the expert can specify an argument that confirms the class value of the example. If the system receives the answer "don't know", it stops the process and tries to find another critical example.

Step 3b: Add arguments. The argument is usually given in natural language and must be translated into domain description language (attributes). One argument supports its allegation with a number of reasons. The role of the expert is also to introduce new concepts to the domain. These concepts are added as new attributes so that they can appear in an argument attached to the example.

Step 3c: Discover counter-examples. A *counter-example* is an example from the opposite class that is consistent with the expert's argument.

Step 3d: Improve arguments. The expert must revise the first argument in relation to the counter-example. This step is similar to steps 1 and 2 with one essential difference; the expert is now asked: "Why is the critical example in one class and why the counter-example in the other?" Note that the argument is always attached to the critical example (and never to the counter-example).

Step 3e: Return to step 3c when a counter-example is found.

In the context of argument-based intelligent tutoring systems, the ABML Knowledge Refinement Loop is used to obtain expert concepts in the form of attributes that describe the domain well, are suitable for teaching and also enable successful rule-based models that achieve high predictive accuracy in the given domain.

3 Arguing to Learn

The ABML Knowledge Refinement Loop can also be used to create meaningful learning experience for a student (see Fig. 2). To achieve this, the intelligent

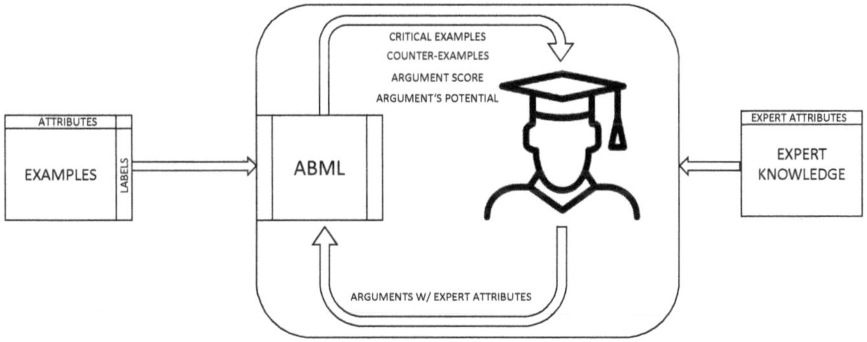

Fig. 2. ABML knowledge refinement loop with the student.

tutoring system must be able to do more than just automatically select relevant examples and counter-examples to be explained by the students. Namely, there should also be automated feedback on the students' arguments, including the generation of hints. The role of an argument-based intelligent tutoring system is then to train students to find the most relevant arguments. Students are encouraged to formulate their arguments using concepts presented by the expert. The students get to know the high-level domain concepts and then use them to argue about automatically selected examples.

We describe three types of feedback on students' arguments, all of which are automatically generated by the underlying machine learning algorithm to help the students construct better arguments and therefore learn faster.

3.1 Counter-Examples

The feedback comes in three forms. The first are the counter-examples that are already inherent in the ABML process. A counter-example is an instance from the data that is consistent with the reasons in the given argument, but whose class value is different from the conclusion of the argument. Therefore, the counter-example is a direct rebuttal to the student's argument. The student must either revise the original argument or accept the counter-example as an exception.

In contrast to earlier applications of the ABML Knowledge Refinement Loop (e.g. [15]), our implementation allows the simultaneous comparison of the critical example with several counter-examples. We believe that this approach allows the student to argue better, as some of the counter-examples are less relevant than others.

3.2 Assessment of the Quality of the Argument

The second type of feedback is an assessment of the *quality* of the argument. A good argument gives reasons for decisions that distinguish the critical example from examples from another class. A possible formula for estimating the quality could therefore be to simply count the number of counter-examples: An argument without counter arguments is generally considered to be a strong argument. However, this method considers very specific arguments (e.g. arguments that only apply to the critical example) to be good. Such specific knowledge is rarely required, we usually prefer general knowledge, which can be applied in several cases.

Therefore, we propose to use the *m-estimate* of probability to estimate the quality of an argument. The formula of the m-estimate balances between the prior probability and the probability assessed from the data:

$$Q(a) = \frac{p + m \cdot p_a}{p + n + m}. \tag{1}$$

Here, p is the number of all covered instances that have the same class value as the critical example, and n is the number of all data instances of another class covered by the argument. We say that an argument covers an instance if the reasons of the argument are consistent with the feature values of the instance. The prior probability p_a and the value m are the parameters of the method used to control how general arguments should be. We estimated the prior probability p_a from the data and set m to 2.

Consider, for example, the argument given to the following critical example:

CREDIT SCORE is *good* because EQUITY RATIO is *high*.

The student stated that this company has a good credit score, as its equity ratio (the proportion of equity in the company's assets) is high. Before the method can evaluate such an argument, it must first determine the threshold value for the label "high". With the entropy-based discretization method, the best threshold for our data was about 40, hence the grounded argument is:

CREDIT SCORE is *good* because EQUITY RATIO $>$ 40 (51, 14).

The values 51 and 14 in brackets correspond to the values p and n, respectively. The estimated quality of this argument using the m-estimate is thus 0.77.

3.3 Potential of the Argument

The last and third type of feedback is the *potential* of the argument. After the student has received an estimate of the quality of his argument, we also give him an estimate of how much the quality would increase if he had improved the argument.

The quality of an argument can be improved either by removing some of the reasons or by adding new reasons. In the first case, we search the existing reasons and evaluate the argument at each step without this reason. For the latter option, we attach the student's argument to the critical example in the data and use the ABCN2 algorithm to induce a set of rules consistent with that argument (this is the same as Steps 3 and 1 in the knowledge refinement loop). The highest estimated quality (of pruned and induced rules) is the potential of the argument provided.

For example, suppose the student has improved his previous argument by adding a new reason:

CREDIT SCORE is *good* because EQUITY RATIO is *high* and CURRENT RATIO is *high*.

The quality of this argument is 0.84. With the ABML method we can induce several classification rules containing EQUITY RATIO and CURRENT RATIO in their condition parts. The most accurate one was:

if NET INCOME $>$ €122,640 and EQUITY RATIO $>$ 30 and CURRENT RATIO $>$ 0.85 then CREDIT SCORE is *high*.

The classification accuracy (estimated with m-estimate, same parameters as above) of the rule is 0.98. This is also the potential of the above argument, since the quality of the best pruned argument is lower (0.77). The potential tells the student that his argument can be improved from 0.84 to 0.98.

4 Case Study

4.1 Domain Description

Credit risk assessment plays an important role in ensuring the financial health of financial and non-financial institutions. Based on a *credit score*, the lender determines whether the company is suitable for lending and how high the price should be. The credit scores are assigned to companies on the basis of their annual financial statements. Arguing what constitutes the credit score of a particular company can significantly improve the understanding of the financial statements [6].

For the machine learning problem, we distinguished between companies with good credit scores and those with bad credit scores. We obtained annual financial statements and credit scores for 325 Slovenian companies from an institution specialising in issuing credit scores. The annual financial statements show the company's business activities in the previous year and are calculated once a year. There were 180 examples of companies with a *good* score and 145 companies with a *bad* score.

At the beginning of the machine learning process, the domain expert selected 25 features (attributes) describing each company. Of these, 9 were from the Income Statement (net sales, cost of goods and services, cost of labor, depreciation, financial expenses, interest, EBIT, EBITDA, net income), 11 from the Balance Sheet (assets, equity, debt, cash, long-term assets, short-term assets, total operating liabilities, short-term operating liabilities, long-term liabilities, short-term liabilities, inventories), 2 from the Cash Flow Statement (FFO - fund from operations, OCF - operating cash flow), and the remaining 3 were general descriptive attributes (activity, size, ownership type).

4.2 Knowledge Elicitation from the Financial Expert

The goal of the knowledge elicitation from the expert is (1) to obtain a rule-based model consistent with his knowledge, and (2) to obtain relevant description language in the form of new features that would describe the domain well and are suitable for teaching. In the present case study, the knowledge elicitation process consisted of 10 iterations. The financial expert introduced 9 new attributes during the process. The new attributes also contributed to a more successful rule model: in the interactive sessions with students (see Sect. 4.3), using the expert's attributes in arguments lead to classification accuracies up to 97%.

The target concepts obtained in the knowledge elicitation process were: Debt to Total Assets Ratio, Current Ratio, Long-Term Sales Growth Rate, Short-Term Sales Growth Rate, EBIT Margin Change, Net Debt To EBITDA Ratio, Equity Ratio, TIE - Times Interest Earned, ROA - Return on Assets. An interested reader can find descriptions of these concepts in introductory books to financial accounting such as [10].

4.3 Interactive Learning Session

In the learning session, each student looks at the annual financial statements of automatically selected companies and faces the challenge of arguing whether a particular company has a good or bad credit score. Students are instructed to use in their arguments the expert features obtained through the process of knowledge elicitation described in Sect. 4.2. This means that the goal of the interaction is that the student is able to explain the creditworthiness of a company using the expressive language of the expert. We will now describe an iteration of one of the learning sessions.

The financial statement and the value of the expert attributes of the training example B.132 were shown to the student (see left part of Fig. 3). The student's task was to explain why this company has a bad credit score.

The student's explanation was: "The company has a certain profit (looking at EBIT and EBITDA in the financial statement) and sales are growing, but also has significant liabilities. Only a small part of the total assets is financed by stockholders, as opposed to creditors. The main reason for the bad credit score seems to be low Equity Ratio." The student also mentioned a high Net Debt To EBITDA Ratio value – note that this attribute introduced by the financial expert is one of the target concepts that the student has to master – but was nevertheless satisfied with the following argument:

CREDIT SCORE is *bad* because EQUITY RATIO is *low*

New rules were obtained and the system identified several counter-examples, such as the one shown in the right side of Fig. 3. The estimated quality of the student's argument was 0.89, which means that the student actually gave a rather good reason why the credit score of the critical example is bad, but, as implied by the potential, can still be significantly improved. The counter-example shown in Fig. 3 also has a low Equity Ratio, but comes from the opposite class. That is, it has a good credit score *despite* low Equity Ratio.

The user noticed a very big difference in the values of Times Interest Earned (TIE). He looked up the definition (available in the application) and found that this attribute indicates how many times a company's earnings cover its interest payments and indicates the probability that a company is (not) able to meet its interest payment obligations. If the student had clicked on "Hints" button in the application, he would get TIE among the possible options to improve the argument. However, he would still have to find out for himself whether a high or low value of TIE indicates good or bad credit score. The student decided to extend the argument:

CREDIT SCORE is *bad* because EQUITY RATIO is *low* and TIE is *low*

This time the potential of the argument implied that the argument could not be further improved, therefore the student demanded a new critical example.

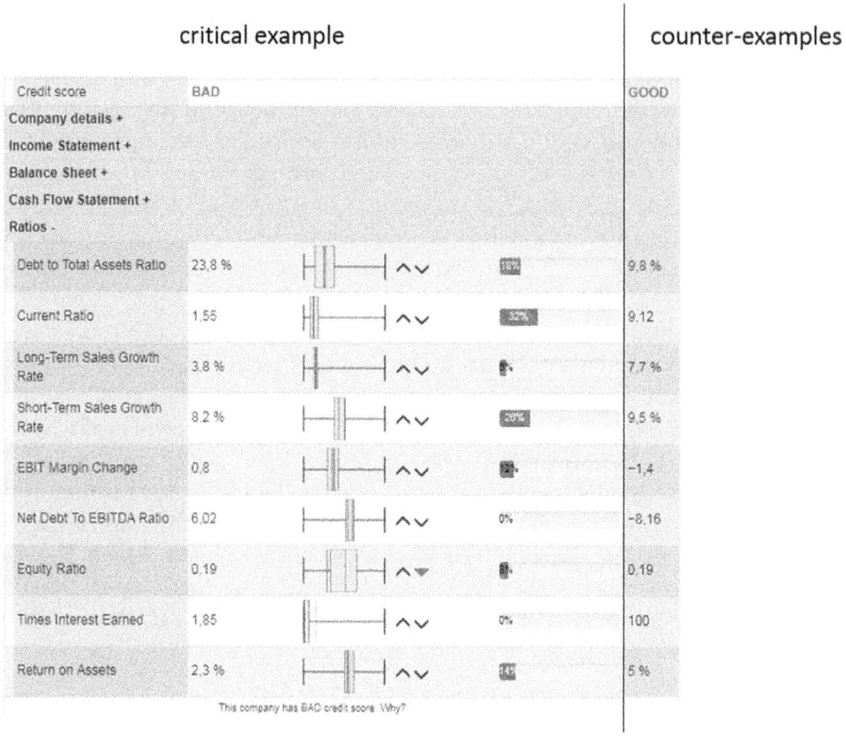

Fig. 3. The student was asked to improve the argument. The attribute values of the critical example are shown on the left and the values of a counter-example are on the right.

5 Assessment

The main objective of the experiments presented in this section was to observe whether automatically generated feedback can enable meaningful interaction suitable for learning. The authors in [1] point out that most studies associated with arguing to learn report great individual differences in the use of tools. They suggest that care must be taken not to fall into the trap of deciding too early which use is correct and which one is not. Since there are many possible solutions and applications of our learning approach through argumentation with the argument-based machine learning framework, we see the results of the pilot experiments described in the sequel of this chapter more as a proof of concept than as a precise evaluation of the concrete applications.

We conducted a pilot experiment with three students. It consisted of 31 iterations such as the one described in Sect. 4.3. All students started the interactive learning session with the same data. The average time per session was 2.83 h. The average number of arguments analyzed was 2.49 (s = 0.37) per iteration. The student typically analyzed more than one argument per iteration, since, with the help of the automatically generated feedback, the arguments were refined several times.

To assess the students' learning performance, we asked them in a post-test to assign credit scores to 30 previously unseen examples. The students' classification accuracy was 87%. We see this as a very positive result, considering that these students had a rather poor understanding of financial statements only a couple of hours earlier and were not aware of the high-level concepts reflected in the expert attributes.

In another experiment, nine students took an online pre-test in which they were asked to tell the credit score of five companies, provide arguments for their decisions and express their confidence when making the decision on a scale from 1 (low) to 5 (high). Then they used the online tool (see Fig. 3) for about 45 min. In a subsequent online post-test, the students were asked exactly the same questions as in the pre-test. We found that the estimated quality of the students' arguments increased on average from 0.85 to 0.87, while the confidence increased on average from 2.70 to 3.58. It appeared that at the end of the process the students could confidently use the high-level concepts introduced by the financial expert in their arguments.

6 Conclusions

We introduced the core mechanisms behind a novel kind of argument-based intelligent tutoring systems based on argument-based machine learning. The main mechanism enabling an argument-based interactive learning session between the student and the computer is called *argument-based machine learning knowledge refinement loop*. By using a machine learning algorithm capable of taking into account a student's arguments, the system automatically selects relevant examples and counter-examples to be explained by the student. The role of an argument-based intelligent tutoring system is to train students to find the most relevant arguments. The students get to know the high-level domain concepts and use them to argue about automatically selected examples. In the interactive learning session, they receive three types of automatically generated feedback: (1) a set of counter-examples, (2) a numerical evaluation of the quality of the argument, and (3) the potential of the argument or how to extend the argument to make it more effective.

The beauty of this approach to developing intelligent tutoring systems is that, at least in principle, any domain that can be successfully tackled by supervised machine learning can be taught in an interactive learning environment that is able to automatically select relevant examples and counter-examples to be explained by the students. To this end, as a line of future work, we are considering the implementation of a multi-domain online learning platform based on argument-based machine learning, taking into account the design principles of successful intelligent tutoring systems [14].

References

1. Andriessen, J., Baker, M.: Arguing to learn. In: Sawyer, R.K. (ed.) The Cambridge Handbook of the Learning Sciences, chap. 22, pp. 439–460. Cambridge Handbooks in Psychology, Cambridge University Press (2014)
2. Andriessen, J., Baker, M., Suthers, D.D.: Arguing to learn: confronting cognitions in computer-supported collaborative learning environments, vol. 1. Springer, Heidelberg (2013). https://doi.org/10.1007/978-94-017-0781-7
3. Bransford, J.D., Brown, A., Cocking, R.: How People Learn: Mind, Brain, Experience, and School. National Research Council, Washington (1999)
4. Chi, M.T., VanLehn, K.A.: The content of physics self-explanations. J. Learn. Sci. **1**(1), 69–105 (1991)
5. Clark, P., Boswell, R.: Rule induction with CN2: some recent improvements. In: Kodratoff, Y. (ed.) EWSL 1991. LNCS, vol. 482, pp. 151–163. Springer, Heidelberg (1991). https://doi.org/10.1007/BFb0017011
6. Ganguin, B., Bilardello, J.: Standard and Poor's Fundamentals of Corporate Credit Analysis. McGraw-Hill Professional Publishing, New York (2004)
7. Groznik, V., et al.: Elicitation of neurological knowledge with argument-based machine learning. Artif. Intell. Med. **57**(2), 133–144 (2013)
8. Guid, M., et al.: ABML knowledge refinement loop: a case study. In: Chen, L., Felfernig, A., Liu, J., Raś, Z.W. (eds.) ISMIS 2012. LNCS (LNAI), vol. 7661, pp. 41–50. Springer, Heidelberg (2012). https://doi.org/10.1007/978-3-642-34624-8_5
9. Guid, M., Možina, M., Krivec, J., Sadikov, A., Bratko, I.: Learning positional features for annotating chess games: a case study. In: van den Herik, H.J., Xu, X., Ma, Z., Winands, M.H.M. (eds.) CG 2008. LNCS, vol. 5131, pp. 192–204. Springer, Heidelberg (2008). https://doi.org/10.1007/978-3-540-87608-3_18
10. Holt, R.: Financial Accounting: A Management Perspective. Ivy Learning Systems (2001)
11. Možina, M., Guid, M., Krivec, J., Sadikov, A., Bratko, I.: Fighting knowledge acquisition bottleneck with argument based machine learning. In: The 18th European Conference on Artificial Intelligence (ECAI), pp. 234–238. Patras, Greece (2008)
12. Možina, M., Žabkar, J., Bratko, I.: Argument based machine learning. Artif. Intell. **171**(10/15), 922–937 (2007)
13. Možina, M., Lazar, T., Bratko, I.: Identifying typical approaches and errors in prolog programming with argument-based machine learning. Expert Syst. Appl. **112**, 110–124 (2018)
14. Woolf, B.P.: Building Intelligent Interactive Tutors: Student-centered strategies for revolutionizing E-learning. Morgan Kaufmann Publishers Inc., San Francisco (2008)
15. Zapušek, M., Možina, M., Bratko, I., Rugelj, J., Guid, M.: Designing an interactive teaching tool with ABML knowledge refinement loop. In: Trausan-Matu, S., Boyer, K.E., Crosby, M., Panourgia, K. (eds.) ITS 2014. LNCS, vol. 8474, pp. 575–582. Springer, Cham (2014). https://doi.org/10.1007/978-3-319-07221-0_73

Model for Data Analysis Process and Its Relationship to the Hypothesis-Driven and Data-Driven Research Approaches

Miki Matsumuro[1(✉)] and Kazuhisa Miwa[2]

[1] College of Information Science and Engineering, Ritsumeikan University, 1-1-1 Noji-higashi, Kusatsu, Shiga, Japan
matumuro@rm.is.ritsumei.ac.jp

[2] Graduate School of Informatics, Nagoya University, Furo-cho, Chikusa-ku, Nagoya, Japan

Abstract. We propose a model explaining a process of the data analysis in the form of the dual space search: data space and hypothesis space. Based on our model, we developed two hypotheses about the relationship between the search in the data space and two scientific research approaches; hypothesis-driven approach and data-driven approach. Generating a testable hypothesis before an analysis (hypothesis-driven) would facilitate the detailed analyses of the variables related to the hypothesis but restrict a search in the data space. On the other hand, the data analysis without a concrete hypothesis (data-driven) facilitates the superficial but broad search in the data space. The results of our experiment using two kinds of the analysis-support system supported these two hypotheses. Our model could successfully explain the process of data analysis and will help design a learning environment or a support system for data analysis.

Keywords: Scientific research process · Research approach · Data analysis · Hypothesis-driven · Data-driven

1 Introduction

In the research process, we conduct an experiment and collect values for many types of variables. For example, when we investigate the learning process of students, we give them problems and measure the number of solved problems, a time to solve a problem, pre- and post-test scores, answers to a questionnaire, and so on. Most of the previous studies have focused on such process primarily; namely, planning and conducting the experiment.

However, to draw valuable conclusions, the data analysis process after the experiment is also crucial. The process includes selecting the variables to be analyzed from the measured variables and the analysis method to be applied. We focus on such data analysis process after conducting the experiment; especially, on selecting the variables to be analyzed. The purpose of this study is to propose a model explaining the data analysis process and confirm two hypotheses derived from the model using two analysis-support systems.

A. Coy et al. (Eds.): ITS 2019, LNCS 11528, pp. 123–132, 2019.
https://doi.org/10.1007/978-3-030-22244-4_16

2 Dual Space Search for Data Analysis Process

An orientation of what stage the instant study is at in the process of conducting a piece of scientific research. One has already conducted his/her planned experiment and collected values of many types of variables. The goal now is to derive conclusions by analyzing the data [6].

2.1 Hypothesis Space and Data Space

The scientific discovery process is described as the search in two spaces, hypothesis and experiment spaces [4]. This model mainly described the process of constructing a hypothesis and planning an experiment but does not include the detailed process of data analysis after the researcher conducts experiments. We introduce another space, data space, to explain the data analysis process in the form of the dual space search model; the dual space in this study means data space and hypothesis space shown in Fig. 1.

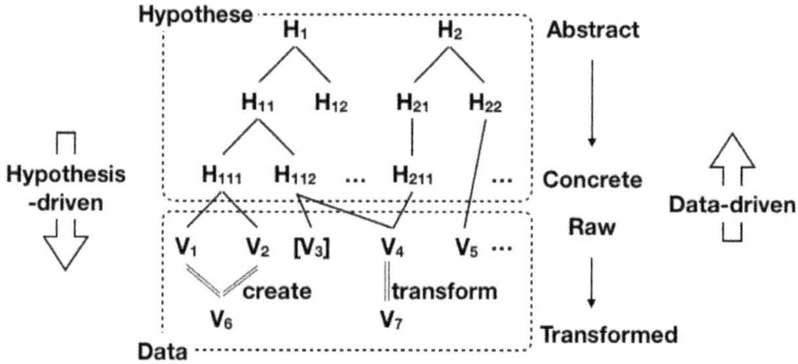

Fig. 1. Hypothesis and data spaces. Hs in the hypothesis space mean hypotheses. Vs in the data space mean variables.

2.2 Data Space

The data space contains all the collected data in the experiment and related data. The data is defined as the values for variables and each variable is a candidate to be a dependent variable [1]. A search was conducted in the data space via data analysis [2].

Particular variables have been selected as the dependent variables and their values analyzed by statistical analysis. Only those selected as dependent variables are marked as searched. When the researchers analyze an identical variable or similar variables repeatedly, their search in the data space is biased.

The researchers are able to create a new variable, which cannot be measured directly, by combining the values of other variables [5]. For example, the increase of a certain test score is created by subtracting the pre-test score from the post-test score.

2.3 Hypothesis Space

The hypothesis space contains all hypotheses, which have a hierarchical structure from an abstract to a concrete level [9]. The hypotheses of an abstract level are general and explain phenomena in a broad scope but cannot make a certain prediction of a related data trend. Hypotheses at a concrete level are strict and explain only a few phenomena, but can make a certain prediction.

The researchers select a hypothesis to be tested [4, 9]. After they focused on a certain middle- or high-level abstraction hypothesis, they select hypotheses from its children, which restrict the search in the hypothesis space. The sufficiently concrete hypothesis was evaluated through value analysis of the corresponding variables. When the results of the analyses are consistent with the prediction from the hypothesis, the hypothesis is supported.

2.4 Search Pattern

Individual differences and the effects of the analysis approach or strategy are represented by the different search pattern in the data and hypothesis spaces. Our model also predicts the behavior of researchers using a certain approach. In this part, we discuss the two approaches in the form of our dual space search model and predict the behaviors during the analysis process.

Hypothesis-Driven Approach. In the hypothesis-driven approach, scientific researches conducted based on a certain hypothesis [4, 7–9]. They generate a testable hypothesis and then manipulate the independent variables and measure the values of the dependent variables, both of which reflect what is stated in the hypothesis. This approach is recognized as the fundamental one for scientific progress.

In this approach, the two spaces are searched from the hypothesis space to the data space. Researchers using the hypothesis-driven approach have their hypothesis, usually a middle-level hypothesis. Therefore, only variables corresponding to the children of their hypothesis are analyzed. In short, because the search in the hypothesis space is restricted, the search in the data space is also restricted.

If the participants have their hypothesis, they try to acquire as much evidence as they can, which support their hypothesis. Additionally, they can predict what types of variables are needed to confirm their hypothesis. Therefore, if they cannot find the variables they need, they try to create them.

Data-Driven Approach. The data-driven approach, wherein conclusions are derived from data analysis with no concrete hypothesis, has attracted a lot of attention recently. In this approach, researchers start with an abstract purpose or research question. The collected data is analyzed to specify the relationship between the variables without a concrete hypothesis [2, 3]. The hypotheses are derived from the results of the analyses; therefore, the two spaces are searched from the data space to the hypothesis space.

Researchers select dependent variables based on their purpose, interest, and sometimes intuition. All variables in the data space have the possibility to be selected as dependent variables, so the data space is not restricted.

In the data-driven approach, researchers just try every variable they are interested in, while in the hypothesis-driven approach, only variables that support their hypothesis are selected. The need to create a new variable is small for them. Therefore, they analyze the collected data without creating any new variables.

3 Hypotheses in This Study

We tested two hypotheses about a relationship between each approach and search in the data space developed based on our model.

Depth Hypothesis. The hypothesis-driven approach facilitates a deeper search in the data space, namely creating new variables, compared to the data-driven approach.

Width Hypothesis. The data-driven approach facilitates a wider search in the data space compared to the hypothesis-driven approach.

4 Analysis-Support System

We developed two kinds of analysis-support systems as shown in Fig. 2, one of which was provided to the participants instructed to use the hypothesis-driven approach (H-driven condition) and another was provided to those instructed to use the data-driven approach (D-driven condition).

Both systems could perform two statistical analyses: a comparison of means (t-test) and a test of correlation for each application condition. All the participants had to do was select the dependent variable(s) from the list. The results of the analyses were presented in a narrative form and as graphs. The participants could create a new variable by selecting two variables and one operator from addition, subtraction, multiplication, and division. They were able to use the new variables for successive analyses.

The analysis process in each condition was controlled by each system using different system prompts. In the H-driven condition, first, the participants were asked to enter the concrete-level hypothesis. Next, they planned an analysis where they entered a variable name(s) they selected as a dependent variable(s) and selected one statistical analysis. Based on their plan, they conducted the statistical analysis. Finally, they concluded whether their hypothesis was supported by the results of the analysis. Then, they repeated these processes cyclically. The system presented their hypothesis in all processes so that they could act based on their hypothesis.

In the D-driven condition, participants started by planning analysis and then conducting it. After that, they interpreted the results of the analysis and described their derived hypothesis of why such results were observed. The format of the planning phase was the same as that used in the H-driven condition. The interpretation phase used the same format as that in the H-driven condition to describe the hypothesis derived in the D-driven condition.

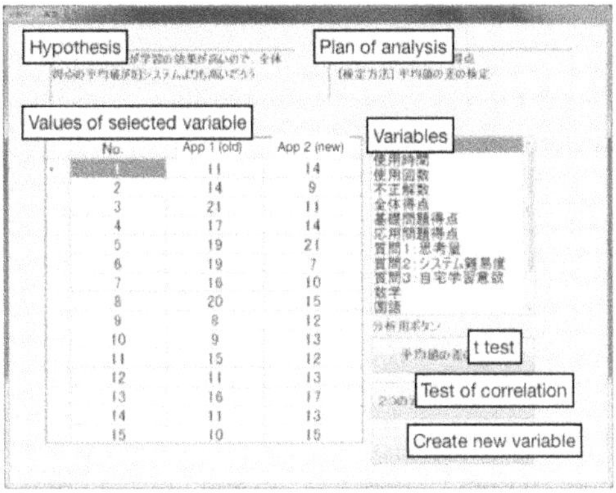

(a) Home screen of the support system.

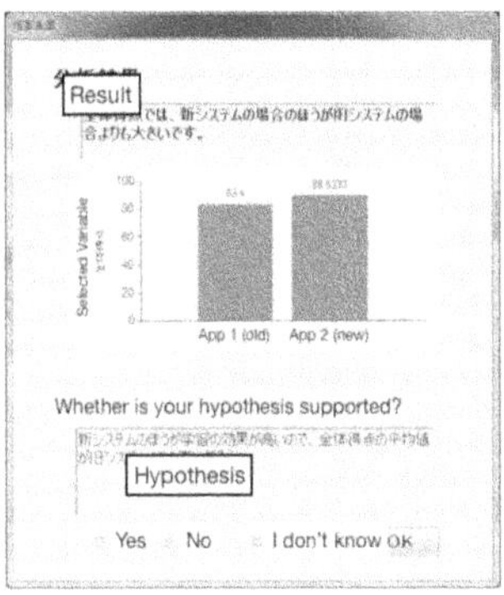

(b) Analysis result screen.

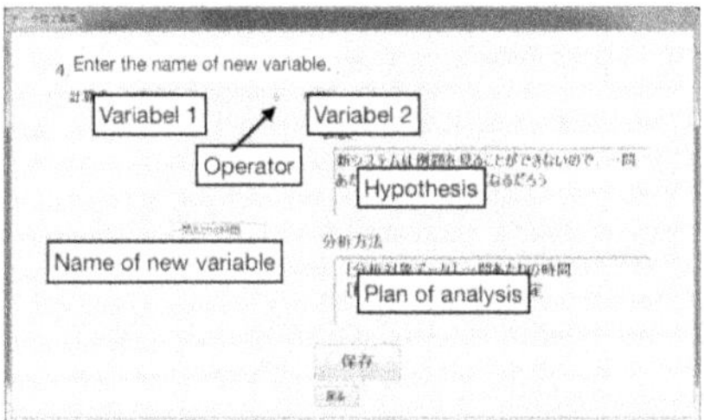

(c) Creation of new variable.

Fig. 2. Example screenshots of the support system for the H-driven condition. The hypothesis window is removed in the D-driven condition.

5 Data Analysis Task

The participants were asked to analyze a data set collected in a fictional experiment. As we described above, they were divided into the H-driven condition or the D-driven condition.

5.1 Scenario

All participants were instructed that: (1) Two electronic applications could be used to complete homework consisting of mathematical proof problems. (2) An evaluation experiment was conducted. (3) Their task was to analyze the data set collected in the evaluation experiment. In the third instruction, the participants were instructed to complete two subtasks. Subtask 1 was to investigate which application offered more learning effects. Subtask 2 was to investigate why one application produced more learning effects more than another.

Participants in the H-driven condition were told that the two homework Apps were named *new App* and *old App*. The role of the participants in the scenario was a professor who developed the *new App*. We gave them the hypothesis: The *new App* would have more learning effects than the old one.

In the D-driven condition, *App 1* and *App 2* were not given any other specific descriptors. The participants did not receive a hypothesis nor were they directed to form one. Rather, they were told that they were to assume they worked at a market research company and were unrelated persons in regard to the application developers.

Homework Apps. Detailed instructions for how to use each application were provided to the participants so that they were able to use the information to interpret the results of their analyses. *App 1 (old)* presented a practice problem next to a worked-out example. In *App 2 (new)*, the worked-out example disappeared before the practice problem was presented.

Procedure for the Evaluation. Experiment The instructed procedure of the fictional evaluation experiment was as follows. Thirty students who had the same level of math ability attended a lecture. Half of them installed *App 1 (old)* and the other half installed *App 2 (new)*. Each student used the provided application at their own pace at home for a week. Then, all students answered a post-test consisting of basic and advanced proof problems and a questionnaire about the applications they had used.

5.2 Data Set

The participants in our experiment were asked to analyze a data set which was comprised of the data collected in the fictional evaluation experiment and other related data. Table 1 shows all the variables that were included in the data set. The variables are divided into five categories.

Values for the Variables. The values of the learning results were determined so as to support the hypothesis for the H-driven condition. The entire and advanced problem scores were significantly higher in the App 2 (new) condition than those in the App 1 (old) condition. We gave a prepared rationale for why App 2 (new) had better teaching effects; "In using App 2 (new), students considered each problem more deeply than when using *App 1 (old)*." To support this rationale, the values of consideration and solving failure were larger in the *App 2 (new)* condition. Additionally, the values of the "time spent per problem" that was acquired by dividing the values of usage time by the number of problems was also larger in the *App 2 (new)* condition. Note that there was no significant difference between the two App conditions for usage time or the number of problems. Each other variable had valid values and reasonable correlations with other variables.

Table 1. Variables in the data set.

Category	Variable name	Collected data
Learning process	Usage time	How long the student used the App
	Number of problems	How many problems were solved using the App
	Number of usages	How many times the student used the App
	Solving failure	How many problems the student failed to solve using the App
Learning result	Entire score	Scores for all problems in the post-test
	Basic score	Scores for basic problems in the post-test
	Advanced score	Scores for advanced problems in the post-test
Answer for questionnaire	Consideration	"Did you consider deeply for each problem?"
	Motivation	"Did you enjoy solving problems using the App?"
	System difficulty	"Did you feel difficulties to use the App?"
Achievement test	Mathematics	Score of achievement test for mathematics
	Japanese	Score of achievement test for Japanese
Lifestyle survey	Number of Bro/Sis	The number of brothers and sisters
	Cellphone use	How many hours the student uses his/her cellphone per day
	After school program	How many hours the student studies other than in the school per week

6 Experiment

6.1 Indices of Depth and Width

Any new variable created by the participants was used as the index of the depth of the search in the data space. As previously described, the new variables meant the search in the area where was not included in the superficial search.

The index of the width of the search was a variance of frequency that the variables in each category used as dependent variables (Shannon's entropy). If the participants searched only in a specific area, the frequency of each category was biased, and entropy was small.

6.2 Method

Participants. Forty-six undergraduate students participated; twenty-three participants were assigned to each of two conditions.

Procedure. First, students attended easy lectures about the two statistical analyses, the procedures for the analyses, and how to use the analysis-support system. After the lectures, the homework applications were introduced along with the fictional evaluation experiment and the data set.

The participants started the analysis task for subtask 1. When they got a satisfying conclusion for it, they entered the conclusion into the system and continued to subtask 2. The analysis task continued for 30 min. Finally, the participants were asked to describe what they concluded about the two applications.

6.3 Results

Two participants who described the findings as inconsistent with the given data were excluded from the following analyses.

Depth of Search. The mean number for the new variables was larger in the H-driven condition than in the D-driven condition (0.851 vs. 0.348, $t(42) = 2.343$, $p = .024$). The results show that the participants in the H-driven condition searched the data space more deeply than those in the D-driven condition did.

Figure 3 shows the categories for the created variables. They were categorized based on the categories of the original variables used for the calculation. If both original variables belonged to the same category, the new variable was also categorized into the same category. If the original variables belonged to different categories, the new variable was categorized into one of them based on its name as decided by the participants. As Fig. 3 shows, the largest number of all new variables was categorized into the process category.

Width of Search. We calculated the frequency that the variables in each category were used for the analyses (Fig. 4). For the new variables, we categorized them into one of the five categories previously described in this paper. The participants in the D-driven condition analyzed the variables in the achievement test and lifestyle survey more frequently.

We calculated the entropy for each condition using the following formula, which was adjusted to a range from 0 to 1. The smaller the entropy number, the more the search was biased.

$$adjusted - H = \left(\sum\nolimits_{C=category} P(C)\log_2 P(C)\right) / \log_2 5 \tag{1}$$

The entropy of the H-driven condition (.708) was significantly smaller than in the D-driven condition (.768; $t(42) = 1.850$, $p = .035$). This result means that in the H-driven condition, the search in the data was biased, especially to the categories directly related the homework applications (i.e., process, result, questionnaire).

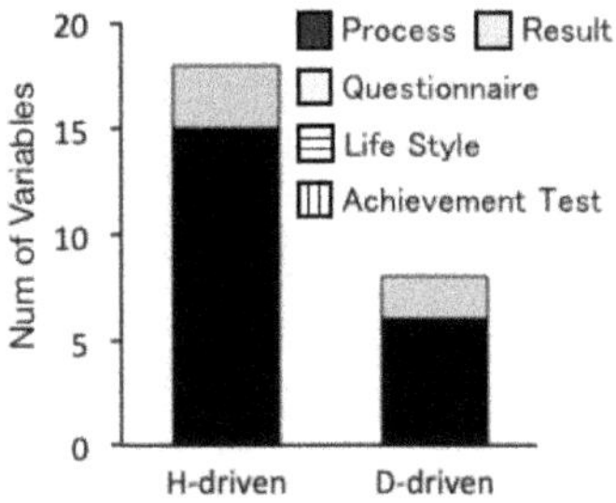

Fig. 3. Types of created variables

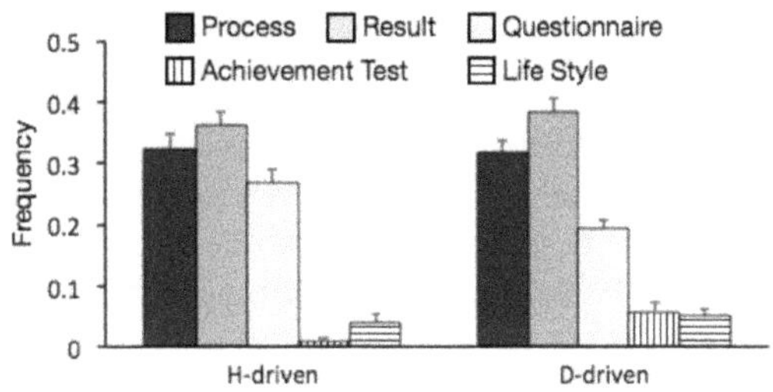

Fig. 4. Frequency of usage as a dependent variable. (bar represents standard error).

7 General Discussion

We proposed the model explaining the data analysis process and tested the following two hypotheses derived from our model.

Depth Hypothesis. The hypothesis-driven approach facilitates a deeper search in the data space compared to the data-driven approach.

Width Hypothesis. The data-driven approach facilitates a wider search in the data space compared to the hypothesis-driven approach.

7.1 Depth of Search

The results demonstrated that the participants in the H-driven condition created more variables, which supports our depth hypothesis: The hypothesis-driven approach facilitates a deeper search in the data space more than the data-driven approach does.

In subtask 2, the participants in the H-driven condition were asked to describe their hypothesis about why the new application had more learning effects. Through the analysis of the answers for the questionnaire, they easily found that the users of new application considered deeply for each problem. This finding led them to the hypothesis that the new application offered more learning effect because of the long consideration. The new variables were created to acquire as much evidence as they could which support this hypothesis. In fact, 8 of the 14 participants who created new variables created a variable "time spent per problem" acquired by dividing the values of usage time by the number of problems named or vice versa.

7.2 Width of Search

As we hypothesized in the width hypothesis, the search in the D-driven condition was broader than in the H-driven condition. When using the data-driven approach, the data space was not restricted; therefore, a broad search was conducted.

The participants in the D-driven condition selected the variables in the category unrelated to the homework application for the analyses, such as the achievement test and lifestyle survey, more than those in the H-driven condition did. The description of what the participants found through the analysis was consistent with their selection. More descriptions about a relationship between the learning effect and the data that was not directly related to the homework application, such as motivation, users' lifestyle, and their family, in the D-driven condition. Some participants derive different conclusions from the results of the same analysis. It may be because they were able to consider the reason for the learning effect flexibly from a broad viewpoint by integrating various results of analyses after the analyses process had finished.

8 Conclusion

Both of two hypotheses based on our model of the data analysis were confirmed by our experiment. The participants who used the data-driven approach conducted broad but shallow searches in the data space. On the other hand, the participants who used the hypothesis-driven approach searched deeply, but only in a restricted data space. Our model, which can successfully explain these differences, will help to design a learning environment or a support system of the data analysis.

Acknowledgements. This work was supported by JSPS KAKENHI Grant Number 18H05320.

References

1. De Mast, J., Trip, A.: Exploratory data analysis in quality-improvement projects. J. Qual. Technol. **39**(4), 301 (2007)
2. Jolaoso, S., Burtner, R., Endert, A.: Toward a deeper understanding of data analysis, sensemaking, and signature discovery. In: Abascal, J., Barbosa, S., Fetter, M., Gross, T., Palanque, P., Winckler, M. (eds.) INTERACT 2015. LNCS, vol. 9297, pp. 463–478. Springer, Cham (2015). https://doi.org/10.1007/978-3-319-22668-2_36
3. Kell, D.B., Oliver, S.G.: Here is the evidence, now what is the hypothesis? the complementary roles of inductive and hypothesis-driven science in the post-genomic era. BioEssays **26**(1), 99–105 (2004)
4. Klahr, D., Dunbar, K.: Dual space search during scientific reasoning. Cogn. Sci. **12**(1), 1–48 (1988)
5. Langley, P.: Data-driven discovery of physical laws. Cogn. Sci. **5**, 31–54 (1981)
6. Pedaste, M., et al.: Phases of inquiry-based learning: definitions and the inquiry cycle. Educ. Res. Rev. **14**, 47–61 (2015)
7. Shute, V.J., Glaser, R.: A large-scale evaluation of an intelligent discovery world: Smithtown. Interact. Learn. Environ. **1**(1), 51–77 (1990)
8. Van Joolingen, W.R., De Jong, T.: Supporting hypothesis generation by learners exploring an interactive computer simulation. Instr. Sci. **20**(5–6), 389–404 (1991)
9. Van Joolingen, W.R., De Jong, T.: An extended dual search space model of scientific discovery learning. Instr. Sci. **25**(5), 307–346 (1997)

On the Discovery of Educational Patterns using Biclustering

Rui Henriques[1](✉), Anna Carolina Finamore[1], and Marco Antonio Casanova[2]

[1] INESC-ID and Instituto Superior Técnico, Universidade de Lisboa, Lisbon, Portugal
rmch@tecnico.ulisboa.pt, anna.couto@ist.utl.pt

[2] PUC-Rio, Rio de Janeiro, Brazil
casanova@inf.puc-rio.br

Abstract. The world-wide drive for academic excellence is placing new requirements on educational data analysis, triggering the need to find less-trivial educational patterns in non-identically distributed data with noise, missing values and non-constant relations. Biclustering, the discovery of a subset of objects (whether students, teachers, researchers, courses and degrees) correlated on a subset of attributes (performance indicators), has unique properties of interest thus being positioned to satisfy the aforementioned needs. Despite its relevance, the potentialities of applying biclustering in the educational domain remain unexplored. This work proposes a structured view on how to apply biclustering to comprehensively explore educational data, with a focus on how to guarantee actionable, robust and statistically significant results. The gathered results from student performance data confirm the relevance of biclustering educational data.

Keywords: Biclustering · Pattern mining · Educational data mining

1 Introduction

Large volumes of educational data are increasingly collected due to a closer monitoring of students, teachers, researchers and staff, with the aim of pursuing academic excellence. This context poses new challenges on extracting meaningful and non-trivial educational patterns to support academic decisions.

Current approaches for educational pattern mining are still unable to reveal the true potential underlying educational data [20]. In its simplest form, educational data gather the performance of a set of objects (such as students, teachers, researchers, courses, degrees, among others) along a set of attributes (performance indicators). Although clustering and pattern mining are typically used to explore such educational data, they are unable to fully extract the hidden knowledge. Clustering simply groups objects (attributes) according to all available values, thus being unable to identify local

R. Henriques and A. C. Finamore—Co-first author

A. Coy et al. (Eds.): ITS 2019, LNCS 11528, pp. 133–144, 2019.
https://doi.org/10.1007/978-3-030-22244-4_17

dependencies (associations on subspaces) and guarantee actionable results. Pattern mining shows limitations on handling numeric or non-identically distributed attributes and lacks robustness to noise and missing data. In addition, it is unable to find non-trivial, yet potentially relevant educational patterns with non-constant coherence, i.e., it cannot consider meaningful variations on the values between objects such as coherent variations on grades from students with different academic performance.

To address the aforementioned limitations, this paper proposes the use of biclustering – subsets of objects meaningfully correlated on a subset of attributes – to comprehensively explore educational data. Although biclustering has been largely used in the biomedical field, its full potential in the educational domain remains untapped.

The results presented in this paper confirm the relevance of biclustering to unravel non-trivial yet meaningful, actionable and statistically significant educational patterns. Specifically, we identify patterns of student performance in topics addressed in a course. Such patterns provide a trustworthy context with enough feedback for the teacher to reform the emphasis given to topics addressed in a course. Our proposal can be extended towards curriculum restructuring; personalized support to students, teachers and researchers; among other ends.

The paper is structured as follows. Section 2 provides the background on biclustering and surveys key contributions from related work. Section 3 describes the unique potentialities of biclustering educational data, and places principles on to achieve them. Section 4 presents results that empirically validate our proposal. Finally, Sect. 5 offers the major concluding remarks.

2 Background

2.1 Biclustering

Definition 1. Given a dataset, $\mathbf{A} = (\mathbf{X}, \mathbf{Y})$, defined by a set of objects $\mathbf{X} = \{\boldsymbol{x}_1, .., \boldsymbol{x}_N\}$, attributes $\mathbf{Y} = \{\boldsymbol{y}_1, .., \boldsymbol{y}_M\}$, and elements $a_{ij} \in \mathbb{R}$ observed in $\boldsymbol{x}_i$ and $\boldsymbol{y}_j$:

- A *bicluster* $\mathbf{B} = (\mathbf{I}, \mathbf{J})$ is a $n \times m$ submatrix of $\mathbf{A}$, where $\mathbf{I} = (i_1, \ldots, i_n) \subset \mathbf{X}$ is a subset of objects and $\mathbf{J} = (j_1, \ldots, j_m) \subset \mathbf{Y}$ is a subset of features;
- The *biclustering task* aims at identifying a set of biclusters $\mathcal{B} = (\mathbf{B}_1, \ldots, \mathbf{B}_s)$ such that each bicluster $\mathbf{B}_k = (\mathbf{I}_k, \mathbf{J}_k)$ satisfies specific *homogeneity*, *dissimilarity* and *statistical significance* criteria.

Homogeneity criteria are commonly guaranteed through the use of a merit function, such as the variance of the values in a bicluster [16]. Merit functions are typically applied to guide the formation of biclusters in greedy and exhaustive searches. In stochastic approaches, a set of parameters that describe the biclustering solution are learned by optimizing a merit (likelihood) function.

The homogeneity criteria determine the structure, coherency and quality of a biclustering solution. The *structure* of a biclustering solution is defined by the number, size, shape and positioning of biclusters. A flexible structure is characterized by an arbitrary number of (possibly overlapping) biclusters. The *coherence* of a bicluster is determined by the observed form of correlation among its elements (coherence assumption) and by the allowed deviations per element against the perfect correlation (coherence strength). The *quality* of a bicluster is defined by the type and amount of accommodated noise. Definitions 2–3 formalize these concepts, and Fig. 1 shows biclusters with different coherence assumptions.

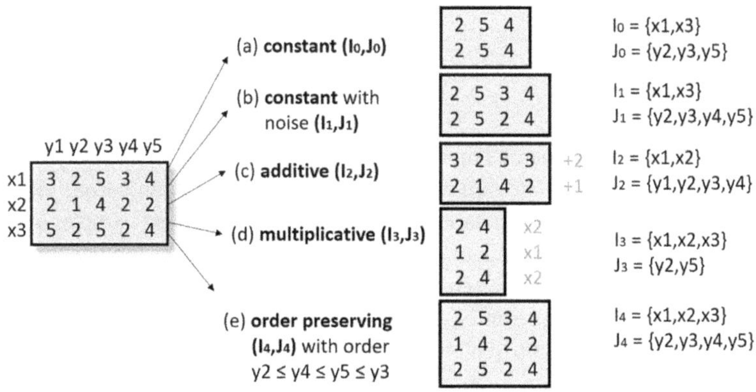

Fig. 1. Discrete biclusters with varying coherence.

Definition 2. Given a numeric dataset $\mathbf{A}$, elements in a bicluster $a_{ij} \in (\mathbf{I}, \mathbf{J})$ have *coherence* across objects iff $a_{ij} = c_j + \gamma_i + \eta_{ij}$ (or attributes iff $a_{ij} = c_i + \gamma_j + \eta_{ij}$), where c_j (or c_i) is the value of attribute $\boldsymbol{y}_j$ (or object $\boldsymbol{x}_i$), γ_i (or γ_j) is the adjustment for object $\boldsymbol{x}_i$ (or attribute $\boldsymbol{y}_j$), and η_{ij} is the noise factor of a_{ij}.

Let $\bar{\mathbf{A}}$ be the amplitude of values in $\mathbf{A}$, *coherence strength* is a value $\delta \in \left[0, \bar{\mathbf{A}}\right]$ such that $a_{ij} = c_j + \gamma_i + \eta_{ij}$ where $\eta_{ij} \in [-\delta/2, \delta/2]$.

Given non-iid finite data where $\boldsymbol{y}_j \in \mathcal{Y}_j$, then $a_{ij} = c_j + \eta_{ij}$ where $a_{ij} \in \mathcal{Y}_j$ and $\delta_j \in \left[0, \bar{\mathcal{Y}}_j\right]$ for continuous attributes and $\delta_j < \left|\mathcal{Y}_j\right|$ for integer attributes.

The γ_i factors define the *coherence assumption*. A bicluster satisfies a *constant* when $\gamma_i = 0$ (or $\gamma_i = 0$), *additive* assumption when $\gamma_i \neq 0$ (or $\gamma_j \neq 0$), and *multiplicative* assumption if a_{ij} is better described by $c_j\gamma_i + \eta_{ij}$ (or $c_i\gamma_j + \eta_{ij}$).

Definition 3. Given a numeric dataset $\mathbf{A}$, a bicluster $(\mathbf{I}, \mathbf{J})$ satisfies the *order-preserving* assumption iff the values for each object in $\mathbf{I}$ (attribute in $\mathbf{J}$) induce the same linear ordering π along the subset of attributes $\mathbf{J}$ (objects $\mathbf{I}$).

Statistical significance criteria, in addition to homogeneity criteria, guarantees that the probability of a bicluster's occurrence (against a null data model) deviates from expectations. **Dissimilarity** criteria can be further placed to comprehensively cover the search space with non-redundant biclusters.

Following Madeira and Oliveira's taxonomy [16], biclustering algorithms can be categorized according to the pursued homogeneity and type of search. Hundreds of biclustering algorithms were proposed in the last decade, as shown by recent surveys [6, 9].

In recent years, a clearer understanding of the synergies between biclustering and pattern mining paved the rise for a new class of algorithms, referred to as pattern-based biclustering algorithms [11]. Pattern-based biclustering algorithms are inherently prepared to efficiently find exhaustive solutions of biclusters and offer the unprecedented possibility to affect their structure, coherency and quality [12, 13]. This behavior explains why this class of biclustering algorithms are receiving an increasing attention in recent years [11]. BicPAMS (Biclustering based on PAttern Mining Software) [12] consistently combines such state-of-the-art contributions on pattern-based biclustering.

2.2 Related Work

Despite the diversity of research contributions on unsupervised educational data mining [2, 8], real-world decisions are still primarily led by data summarization, visualization and statistics. Such approaches are centered on efforts to test simple hypotheses, facilitate searches and support data navigation, whether data is tabular, event-based, relational, multi-dimensional, or semi-structured [8]. In an attempt to automatize educational data analysis and guarantee a focus on less-trivial data relations, contributions in the fields of clustering and pattern mining have been also proposed [2, 7]. In the context of pattern mining, Buldu and Üçgün [4], Chandra and Nandhini [5], Gottin et al. [10], and Olaniyi et al. [17] pursued association rules pertaining to student performance and topic agreement to support curriculum redesign. Sequential pattern mining has been alternatively applied for topic data analysis to model students' behaviors along an educational programme [1, 3]. Results suggest that sequential patterns can be used to enrich training data for improving predictions of students' performance.

Biclustering has been firstly suggested for educational data exploration by Trivedi et al. [18, 19] to understand the impact of tutor interaction in students' performance. To this end, the authors partitioned students and interaction features to produce biclusters for predicting out-of-tutor performance of students. Results show a moderately reduced error (against baseline predictors). Despite its merits, the applied algorithm imposes biclusters to follow a checkboard structure, a severe restriction, which possibly explains the modest results.

Vale et al. [20] offered a comprehensive roadmap on the relevance of biclustering for two distinct sources of educational data: (1) matrices relating students and subjects through achieved marks, where the interest is placed on students showing coherent grades in a particular subset of subjects, and (2) matrices collecting performance indicators of subjects along time with the aim of finding temporal patterns. The goal of the work was to find biclusters not trivially retrieved using alternative pattern mining methods. To this end, xMOTIFs, ISA and OPSM biclustering algorithms are considered. Despite its relevance, the obtained patterns are approximate and not statistically tested.

3 Solution: Biclustering in Educational Data

As surveyed in previous section, pattern-based biclustering approaches provide the unprecedented possibility to comprehensively find patterns in non-iid data with parameterizable homogeneity and guarantees of statistical significance. Despite their relevance, their use to explore educational data remains unassessed. This section provides a structured view on how to bicluster educational data, identifying its unique potentialities.

Real-Valued Educational Patterns. Biclustering seeks patterns in real-valued data with *coherence orientation* along objects or attributes (Definition 2). Illustrating, in student performance analysis, biclusters with patterns on objects reveal coherent grades on specific topics for a subset of students.

Biclustering also allows the calibration of *coherence strength* (Definition 2) – e.g. how much two academic indicators need to differ to be considered dissimilar. Allowing deviations from pattern expectations in real-valued educational data is key to prevent the item-boundaries problem, thus tackling discretization problems faced by classic pattern mining methods. Patterns are inferred from similar (yet non-strictly identical) performance indicators, whether numerical or ordinal.

Comprehensive Educational Data Exploration. Pattern-based biclustering offers principles to find complete solutions of educational patterns by: (1) pursuing multiple homogeneity criteria, including multiple coherence strength thresholds, coherence assumptions and quality thresholds, and (2) exhaustively yet efficiently exploring different regions of the search space, preventing that regions with large patterns jeopardize the search. As a result, less-trivially correlated indicators of academic performance are not neglected. By contrast, classic pattern mining procedures uniquely focus on educational patterns with constant coherence and the underlying searches have efficiency bottlenecks in the presence of lengthy patterns. Furthermore, pattern-based biclustering does not require the input of support thresholds as it explores the search space at different supports, i.e. we do need to place expectations on the minimum number of students/teachers/researchers per pattern. Still, the minimum number of (dissimilar) patterns, minimum percentage of covered data elements, and minimum number of objects and/or performance indicators in a bicluster can be optionally inputted to guide the search. Parameterizable dissimilarity criteria and condensed representations can be placed [12] to prevent redundant educational patterns.

Non-constant Educational Patterns. Depending on the goal, one or more *coherence assumptions* (Definitions 2 and 3) can be pursued. Let us illustrate paradigmatic cases in student performance analysis. The classic constant assumption can be placed to unravel groups of students with similar grades on a subset of topics/courses. However, it is unable to correlate grades from students with different performance profiles. In this context, non-constant patterns can be pursued:

- *additive* pattern: set of students with different average of performance yet coherent grades on a subset of topics explained by shifting factors (Fig. 1c);
- *multiplicative* pattern: set of students with linearly correlated grades on a subset of topics/courses explained by scaling factors (Fig. 1d);
- *order-preserving* pattern: set of students with preserved orderings of grades on a subset of topics/courses (Fig. 1e).

As a result, pattern-based biclustering allows the discovery of less-trivial yet coherent, meaningful and potentially relevant educational relations.

Robustness to Noise and Missing Values. With pattern-based biclustering, and by co trast with classic pattern mining, the user can find biclusters with a parameterizable tolerance to noise. This possibility ensures, for instance, robustness to the inherent subjectivity of Likert scale evaluations in questionnaires.

Similarly, pattern-based biclustering is robust to missing data by permitting the dis covery of biclusters with an upper bound on the allowed amount of missings. This is particularly relevant to handle missing ranks in questionnaire data or missing grades due to unassessed topics or unattended exams.

In turn, this ability to handle missing data allows the discovery of coherent modules (biclusters) in network data (sparse adjacency data) such as student community data or research collaboration data.

Statistical Significance. A sound statistical testing of educational patterns is key to guarantee the absence of spurious relations, validate conclusions inferred from educational patterns, and ensure pattern relevance when building academic reforms and making other decisions. To this end, the statistical tests proposed in BSig [15] are suggested to minimize the number of false positives (output patterns yet not statistically significant) without incurring on false negatives. This is done by approximating a null model of the target educational data and appropriately testing each bicluster in accordance with its underlying coherence.

Other Opportunities. Additional benefits of pattern-based biclustering can be carried towards educational data analysis, including: (1) the removal of uninformative elements in data to guarantee a focus, for instance, on lower student grades or assessments of faculty members suggesting problematic performance; (2) incorporation of domain knowledge to guide the biclustering task, useful in the presence of background data on courses, students or faculty members [14]; and (3) support to classification and regression problems in education in the presence of annotations by guaranteeing the discriminative power of biclusters [11].

4 Results on Student Performance Data

To illustrate the enumerated potentialities of biclustering educational data, we discuss results from student-topic performance data in four major steps. First, we empirically delineate general advantages of biclustering student-topic data. Second, we show that biclustering finds educational patterns robust to noise and missings. Third, we provide evidence for the relevance of finding non-trivial (yet meaningful) educational patterns with non-constant coherence. Finally, we show that biclustering guarantees the statistical significance of relations, providing a trustworthy means for academic reforms.

ADS dataset. The ADS dataset[1] captures the performance of students along the topics of the Advanced Data Structures (ADS) course offered every academic term by the Department of Informatics of the Pontifical Catholic University of Rio de Janeiro (PUC-Rio). The dataset combines the results of exams, covering 10 academic terms in

[1] BicPAMS available at https://web.ist.utl.pt/rmch/bicpams/. ADS data available upon request.

which the course was under the responsibility of the same teacher, amounting a total of 229 students and 325 enrollments.

Experimental setting. The BicPAMS algorithm (See footnote 1) [12] is applied since it consistently integrates the state-of-the-art algorithms on pattern-based biclustering and guarantees the efficiency of the underlying searches. BicPAMS is below used with default parameters: varying coherence strength ($\delta = \bar{A}/|\mathcal{L}|$ where ($|\mathcal{L}| \in \{3, 4, 5\}$), decreasing support until at least 50 dissimilar biclusters are found, up to 30% noisy elements, 0.05 significance level, and a single coherence assumption at a time (constant, additive, multiplicative and order-preserving). Two search iterations were considered by masking the biclusters discovered after the first iteration to ensure a more comprehensive exploration of the data space and a focus on less-trivial educational patterns. Topic-based frequency distributions were approximated, and the statistical tests proposed in [15] were applied to compute the statistical significance of each found bicluster.

4.1 Real-Valued Educational Patterns

Table 1 synthesizes the results produced by biclustering student-topic data with BicPAMS [12]. Confirming the potentialities listed in Sect. 3, BicPAMS was able to efficiently and comprehensively find a large number of homogeneous, dissimilar and statistically significant biclusters – subsets of students with coherent performance on a subset of topics. One can check, for instance, in the first row of Table 1, that among the total number of discovered biclusters (135), we found that 120 of them are statistically significant with a p-value lower that 0.1%. Given these 135 biclusters, there are approximately $u(|\mathbf{I}_1|, \ldots, |\mathbf{I}_{135}|) = 16$ students per bicluster on average and $u(|\mathbf{J}_1|, \ldots, |\mathbf{J}_{135}|) = 3$ topics per bicluster on average considering a constant assumption, three bins ($|\mathcal{L}| = 3$ and $\delta = \bar{\mathbf{A}}/|\mathcal{L}|$), and a perfect quality (100%/no noise).

Table 1. Properties of the biclustering solutions found in ADS data with BicPAMS when varying the homogeneity criteria.

Assumption	$\|L\|$	Quality	#bics	$\mu(\|I\|) \pm\sigma(\|I\|)$	$\mu(\|J\|) \pm\sigma(\|J\|)$	p-value >0.01	p-value $\in$ [0.1, 1E−3]	p-value <1E−3
Constant	3	100%	135	15.6 ± 5.8	2.9 ± 0.4	10	15	120
Constant	3	70%	123	20.3 ± 5.1	3.1 ± 0.3	2	10	111
Constant	4	70%	168	15.1 ± 4.8	3.0 ± 0.1	10	26	132
Constant	5	70%	241	10.8 ± 3.7	3.1 ± 0.2	9	65	167
Additive	4	70%	310	15.3 ± 8.2	3.1 ± 0.4	17	23	270
Multiplicative	4	70%	195	14.3 ± 5.3	3.1 ± 0.4	9	13	173
Order-preserving	_	70%	91	27.4 ± 4.4	3.4 ± 0.5	11	20	60

These initial results further show the impact of tolerating noise by placing different coherence assumptions (such as the order-preserving assumption), and parameterizing coherence strength ($\delta \propto 1/|\mathcal{L}|$) on the biclustering solution.

4.2 Constant Educational Patterns

Table 2 provides the details of an illustrative set of four constant biclusters (and the respective performance pattern, subset of topics, coherence strength and statistical significance) using BicPAMS with default parameters. Each bicluster shows a unique pattern of performance. For instance, bicluster $\mathbf{B}_5$ reveals a group of 13 students who coherently encountered moderate, delineate and no difficulties (corresponding to the pattern $\{1, 0, 4\}$ using 5 bins where 0 denotes a low grade and 4 an excelling grade) in 3 topics (binary searches, bit-vectors and complexity).

Figure 2a visually depicts an additional constant bicluster. Each line in the chart represents a student and her/his grades on along the 3 topics in the bicluster.

These results motivate the relevance of finding constant biclusters to understand coherent patterns of difficulty between topics for a statistically significant population of students. One can check that a bicluster considers both identical grades (where lines converge) and more loosely similar values (where lines diverge). The profile of the students in a specific bicluster can be further analyzed to further understand its influence on the resulting performance.

Table 2. Constant biclusters found in ADS data.

Bicluster properties	Pattern (studentID⇒values)
$\mathbf{B}_1$ topics=[Bitvector,Heap, Dijkstra,SocialNet] #students=16, $\lvert L\rvert$=3 p-value=**1.06E-18**	x_{14} 0 0 0 0 x_{21} 0 0 0 0 ... x_{267} 0 1 0 0 x_{317} 0 0 0 0
$\mathbf{B}_2$ topics =[B-TreesRemoval, Bitvector,Complexity] #students=16, $\lvert L\rvert$=4 p-value=**8.7E-4**	x_{11} 0 0 3 x_{55} 0 0 3 ... x_{269} 0 0 3 x_{317} 0 0 3
$\mathbf{B}_3$ topics=[B-TreesRemoval, Heap,Morton] #students=14, $\lvert L\rvert$=4 p-value=**1.6E-8**	x_{15} 0 0 1 x_{23} 0 0 1 ... x_{280} 0 0 1 x_{320} 0 0 1
$\mathbf{B}_4$ topics=[Hash,Union-Find, Complexity] #students=15, $\lvert L\rvert$=4 p-value=**1.3E-6**	x_{11} 1 3 3 x_{16} 1 3 3 ... x_{292} 1 3 3 x_{321} 1 3 3
$\mathbf{B}_5$ topics=[BinarySearchT, Bitvector,Complexity] #students=13, $\lvert L\rvert$=5 p-value=**6E-7**	x_{11} 1 0 4 x_{46} 1 0 4 ... x_{292} 1 0 4 x_{299} 1 0 4

Table 3. Order-preserving biclusters in ADS data.

Bicluster properties	Pattern (studentID⇒values)
$\mathbf{B}_6$ topics=[B-TreesRemoval, BinarySearchTrees,Heap] Order of difficulty: Heap<BinarySearchT< <B-TreesRemoval #students=35	x_{42} 9 7 0 x_{103} 7 4 3 x_{181} 5 3 2 x_{218} 4 3 2 x_{247} 7 6 4 x_{256} 7 4 1 x_{260} 9 8 4 x_{265} 9 3 2 x_{281} 9 8 4 x_{286} 7 6 2 x_{292} 8 3 2 x_{307} 9 8 7 x_{308} 9 3 0 ...
$\mathbf{B}_7$ topics=[B-TreesInsertion, Bitvector,Hash,Heap] Order of difficulty: Heap<B-TreesInsertion= =Bitvector<Hash #students=17	x_{12} 8 8 3 9 x_{30} 8 8 0 9 x_{32} 1 1 0 3 x_{94} 8 8 5 9 x_{107} 1 1 0 2 x_{151} 5 5 4 8 x_{267} 1 1 0 4 x_{268} 1 1 0 9 x_{269} 1 1 0 7 x_{289} 8 8 0 9 x_{310} 1 1 0 9 x_{313} 8 8 7 9 ...

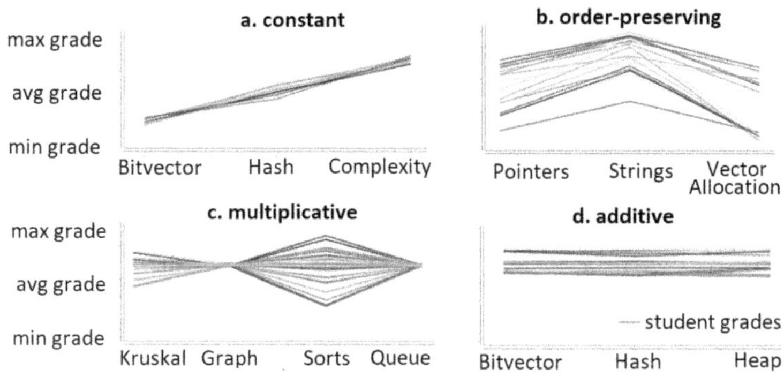

Fig. 2. Set of (a) constant, (b) order-preserving, (c) multiplicative, and (d) additive biclusters found in ADS data (subsets of students with coherent grades on subsets of topics in the absence and presence of ordering, scaling and shifting factors).

A closer analysis of the found biclusters shows their robustness to the item-boundaries problem: students with slightly deviating grades from pattern expectations are not excluded from the bicluster. This allows the analysis of real-valued or/and integer data without the drawbacks of aggregated/discrete views on students' performance.

4.3 Non-constant Patterns

Non-constant patterns are suggested if the focus is not on determining levels of performance but to assess the relative difficulty among topics. BicPAMS [12] was applied to find such less-trivial yet relevant topic-student patterns, including patterns with order-preserving, additive, and multiplicative coherence assumptions (Table 1).

Table 3 provides the details of two statistically significant order-preserving biclusters, including the subset of students and topics, and the permutation of topic grades (the pattern). For instance, bicluster $\mathbf{B}_6$ reveals an unexpectedly large group of students with arbitrarily-different grades yet coherently facing more difficulties in heaps, then binary searches and, finally, B-tree removals.

Figure 2 depicts 3 additional biclusters with order-preserving (2b), multiplicative (2c) and additive (2d) coherence. These coherence assumptions are useful to accommodate coherent orders, shifts and scales in student performance, thus being able to account for differences in students' aptitude.

4.4 Noise-Missing Robustness

Tolerance to noise can be customized (see Table 1) in order to comprehensively find patterns with parameterizable degree of quality. In addition to noise-tolerance, η_{ij}, coherence strength $\delta = \bar{\mathbf{A}}/|\mathcal{L}|$ can be explored (Table 1) to comprehensively model relations between students and topics with slight-to-moderate deviations from expectations.

The analysis of the found biclusters further confirms their ability to tolerate missing educational data. ADS data have two major types of missing grades caused by: (1) students not showing up to an exam (not evaluated), and (2) not all topics being covered in the context of an exam applied in a given semester.

4.5 Statistical Significance

Table 1 shows the biclustering ability to find statistically significant relations in student-topic data. A bicluster is statistically significant if the number of students with a given pattern or permutation of topic grades is unexpectedly low [15]. Figure 3 provides a scatter plot of the statistical significance (horizontal axis) and area $|I| \times |J|$ (vertical axis) of constant biclusters with $|\mathcal{L}| = 3$ and >70% quality. This analysis suggests the presence of a soft correlation between size and statistical significance. We observe that few biclusters have low statistical significance (right bottom dots) and therefore should be discarded to not incorrectly bias decisions in educational contexts.

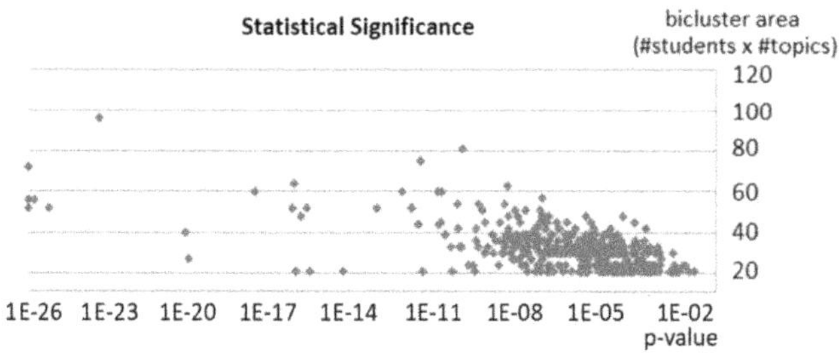

Fig. 3. Statistical significance vs. size of collected constant biclusters ($|\mathcal{L}| = 3$).

5 Conclusions

This work proposed comprehensive principles on how to apply biclustering for the exploration of educational data in order to tackle the limitations of peer unsupervised tasks, such as clustering and pattern mining, and untap the hidden potential underlying educational data by focusing on non-trivial, yet meaningful and statistically significant relations. Pattern-based biclustering searches are suggested to find actionable educational patterns since they hold unique advantages: efficient exploration; optimality guarantees; discovery of non-constant patterns with parameterizable coherence; tolerance to noise and missing data; incorporation of domain knowledge; complete biclustering structures without positioning restrictions; and sound statistical testing.

Results from student-topic data confirm the unique role of biclustering in finding relevant patterns of student performance, such as similar topic difficulties experienced by students with a specific profile (given by constant or additive biclusters) and orders of topic difficulties (given by order-preserving biclusters).

Results further evidence the ability to unveil interpretable patterns with guarantees of statistical significance and robustness, thus providing a trustworthy context with enough feedback for the teacher to reform the emphasis given to topics addressed in a course. A similar analysis can be conducted in alternative educational data domains, including monitored lecturing and research activities.

Acknowledgement. This work is supported by national funds through Fundação para a Ciência e a Tecnologia (FCT) under project iLU DSAIPA/DS/0111/2018 and INESC-ID pluriannual UID/CEC/50021/2019.

References

1. Antunes, C.: Acquiring background knowledge for intelligent tutoring systems. In: EDM (2008)
2. Baker, R.S., Inventado, P.S.: Educational data mining and learning analytics. In: Larusson, J. A., White, B. (eds.) Learning Analytics, pp. 61–75. Springer, New York (2014)
3. Barracosa, J., Antunes, C.: Anticipating teachers performance. In: KDD IW on Knowledge Discovery in Educational Data, pp. 77–82 (2011)
4. Buldu, A., Üçgün, K.: Data mining application on students data. Procedia - Soc. Behav. Sci. **2**(2), 5251–5259 (2010)
5. Chandra, E., Nandhini, K.: Knowledge mining from student data. Eur. J. Sci. Res. **47**(1), 156–163 (2010)
6. Charrad, M., Ben Ahmed, M.: Simultaneous clustering: a survey. In: Kuznetsov, S.O., Mandal, D.P., Kundu, M.K., Pal, S.K. (eds.) PReMI 2011. LNCS, vol. 6744, pp. 370–375. Springer, Heidelberg (2011)
7. Dutt, A., Aghabozrgi, S., Ismail, M.B., Mahroeian, H.: Clustering algorithms applied in educational data mining. IJ Info. Electron. Eng. **5**(2), 112 (2015)
8. Dutt, A., Ismail, M.A., Herawan, T.: A systematic review on educational data mining. IEEE Access **5**, 15991–16005 (2017)
9. Eren, K., Deveci, M., Küçüktunç, O., Çatalyürek, Ü.: A comparative analysis of biclustering algorithms for gene expression data. Brief. Bioinf. **14**(3), 279–292 (2013)
10. Gottin, V., Jiménez, H., Finamore, A.C., Casanova, M.A., Furtado, A.L., Nunes, B.P.: An analysis of degree curricula through mining student records. In: ICALT, pp. 276–280. IEEE (2017)
11. Henriques, R., Antunes, C., Madeira, S.C.: A structured view on pattern mining-based biclustering. Pattern Recognit. **48**(12), 3941–3958 (2015)
12. Henriques, R., Ferreira, F.L., Madeira, S.C.: BicPAMS: software for biological data analysis with pattern-based biclustering. BMC Bioinform. **18**(1), 82 (2017)
13. Henriques, R., Madeira, S.C.: BicPAM: pattern-based biclustering for biomedical data analysis. Algorithms Mol. Biol. **9**(1), 27 (2014)
14. Henriques, R., Madeira, S.C.: BiC2PAM: constraint-guided biclustering for biological data analysis with domain knowledge. Algorithms Mol. Biol. **11**(1), 23 (2016)
15. Henriques, R., Madeira, S.C.: BSig: evaluating the statistical significance of biclustering solutions. Data Min. Knowl. Discov. **32**(1), 124–161 (2018)
16. Madeira, S.C., Oliveira, A.L.: Biclustering algorithms for biological data analysis: a survey. IEEE/ACM Trans. Comput. Biol. Bioinform. **1**(1), 24–45 (2004)
17. Olaniyi, A.S., Abiola, H.M., Taofeekat Tosin, S.I., Kayode, Babatunde, A.N.: Knowledge discovery from educational database using apriori algorithm. CS&Telec. **51**(1) (2017)

18. Trivedi, S., Pardos, Z., Sárkozy, G., Heffernan, N.: Co-clustering by bipartite spectral graph partitioning for out-of-tutor prediction. In: Proceedings of the 5th International Conference on Educational Data Mining, Chania, Greece, 19–21 June 2012, pp. 33–40 (2012)
19. Trivedi, S., Pardos, Z., Sárkozy, G., Heffernan, N.: Spectral clustering in educational data mining. In: EDM (2010)
20. Vale, A., Madeira, S.C., Antunes, C.: Mining coherent evolution patterns in education through biclustering. In: Educational Data Mining (2014)

Parent-Child Interaction in Children's Learning How to Use a New Application

Akihiro Maehigashi(✉) and Sumaru Niida

Interaction Design Laboratory, KDDI Research, Inc.,
2-1-15 Ohara, Fujimino, Saitama, Japan
{ak-maehigashi,niida}@kddi-research.jp

Abstract. This study investigated how parents support their children when they are learning how to use a new application in a situation where the parents do not have any practical knowledge of it and how parental support influences the children's learning. Eleven pairs of parents and children (aged 4–10 years) used a scrapbook calendar application in workshop. We conducted analyses on the application logs, the questionnaire data, and the recorded conversations. The results revealed that there were three different types of groups, the child-led operation group (parents orally supported children based on their prior knowledge and deductions), the parent-led operation group (parents let children participate in operations intermittently), and the leader-led operation group (parents transferred the operation from themselves to children). The children in the child-led and the leader-led operation groups were successful in using the application. Moreover, the children in the child-led operation group were highly motivated to use the application, and in contrast, the children in the parent-led and the leader-led operation groups were less motivated. Based on the results, the application to tutoring systems is discussed.

Keywords: Parent–child interaction · Child-computer interaction · Parental support · Joint media engagement · Motivation

1 Introduction

Due to the development and prevalence of technology, the education, communication, and play of children nowadays are supported by multiple digital devices. In Japan, 37.4% of two-year-old children, 47.5% of three-year-old children, and 50.4% of four-year-old children use any of computers, tablet computers, smart phones and mobile phones [1]. It means that over half of the children aged more than four years use digital devices in their daily lives in Japan. Also, in USA, 9% of under two-year-old children, 27% of two- to four-year-old children, and 37% of five- to eight-year-old children use a mobile device at least once a day [2].

Preschoolers are basically able to use digital devices by observing their parents and siblings using them [3]. Parental attitudes and use of digital devices have a strong influence on their children's use of devices [4]. As children grow older, their independent use of devices increases. When they reach the age of six or more, the influence from parental attitudes toward devices disappears [5]. Although children are able to use

A. Coy et al. (Eds.): ITS 2019, LNCS 11528, pp. 145–155, 2019.
https://doi.org/10.1007/978-3-030-22244-4_18

digital devices by themselves, parents usually participate in their children's use of devices in various ways.

Flynn et al. [6] showed that when children, aged from three to four years, played a computer game for the first time, the parents provided their children with multiple types of support. In particular, parents supported children by telling them what to do (directives), explaining the connection between the game and the screen or the children's experiences (connections), playing instead of children (taking over), giving advice and assistance (motor skill help), or asking children questions about the game or what they were supposed to do (other content-related talk). Moreover, Hinker et al. [7] showed that even when children, aged from four to six years, played a tablet computer game which they were accustomed to play independently, parents provided various types of support. In this case, the parents became involved in the children's activity by playing the game together (teammate), guiding children, providing advice, and asking questions (coach), and observing children's play (speculator). Furthermore, Barron et al. [8] showed that parents also provided various types of support to children even when they were aged from 12 to 14 years. Particularly, parents supported children by teaching them operational methods (teacher), performing with children (project collaborator), seeking to learn opportunities for children (learning broker), providing children with computer-related resources (resource provider), providing information or advice to children on nontechnical issues such as business or artistic design (nontechnical consultant) and so on.

These previous studies indicated that parents actively participate in children's use of digital devices in various ways. Such parent-child joint engagement supports children to learn with, from, and about technology [9]. However, these previous studies did not rigorously discuss the fact that parents are not familiar with all the digital devices. It is common that parents do not have any practical knowledge of new digital devices which children actually use or try to use. In contrast to previous studies, this study focused on parent-child interaction in a situation where children try to use a new application that the parents do not have any practical knowledge of. The purpose of this study was to investigate in what ways parents support children to learn how to use a new application and how the parental support influences learning when parents do not have any practical knowledge of the application.

2 Workshop

2.1 Scrapbook Calendar Application

In the workshop, a scrapbook calendar application [10] was used (Fig. 1a). This application allows users to decorate the calendar with pictures and typed letters or emojis. To decorate the calendar, there were three types of operation, *editing*, *adding*, and *adjusting*. With the editing operation, users could cut pictures from a photo album, draw pictures or letters by tracing the display with a finger, or type letters or emojis with the on-screen keyboard. Using the adding operation, they incorporated the edited materials into the calendar simply by tapping the "done" button displayed in the upper right corner of the calendar. Finally, by means of the adjusting operation, they could change the sizes and the angles of the added materials on the calendar by pinching or rolling the materials with their fingers.

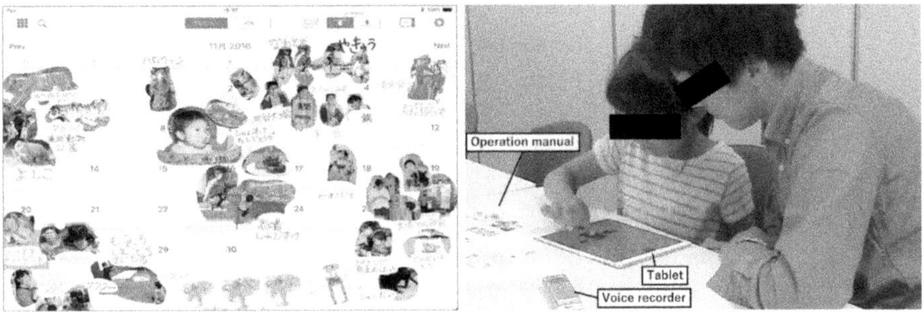

(a) Scrapbook calendar application **(b) Workshop situation**

Fig. 1. (a) Screen capture of the scrapbook calendar application, and (b) workshop situation with parent and child - a still from a video recording.

One of the reasons that we used this application in this study was that it allows a parent and a child to have a range of different interactions. The application can be used by one person or two people. Also, two people can use it alternately, or one person can intervene in the other's operation. The other reason was that it was guaranteed that the participants had never used it before. The application had been newly developed and was yet to be released. Therefore, it could be safely assumed that the participants had not had any previous practical knowledge or experience of the application.

2.2 Method

Participants. Eleven pairs of a parent and a child participated in the workshop. Table 1 shows the summary of the participants. In this study, children aged from 4 to 10 participated in the workshop. Previous studies showed that the children's age strongly influenced their use of digital devices [1, 2]. Therefore, we also focus on how their ability to learn the use of a new application is influenced by their age.

Procedure. The workshop took three hours. A camera and a voice recorder were prepared for each pair to record the participants behaviors and utterances. Figure 1b shows the workshop situation with the parent and child. First, the digital pictures they had brought along for the workshop were transferred to iPads in which the application was already installed. Next, the application was briefly explained to the parents, and an operation manual was also provided. The explanation was made very brief so it was not possible to fully understand how to use the application even with the manual. After that, the first session began. They used the application for about 40 min and decorated the calendar with actual events from the previous one or two months. After the first session, the parents rated the extent to which they agreed or disagreed that their children fully understood the functions and operational methods of the application on a

Table 1. Summary of the participants.

No.	Parent	Age	Child	Age
1	Mother	36	Son	6
2	Father	36	Daughter	8
3	Mother	40	Son	8
4	Father	38	Son	10
5	Father	36	Daughter	5
6	Father	40	Daughter	8
7	Mother	33	Daughter	4
8	Mother	33	Son	6
9	Father	35	Son	7
10	Mother	35	Son	9
11	Father	35	Daughter	9

seven-point scale (1: strongly disagree - 7: strongly agree). After a 15-minute break, the second session began. In the second session, they used the application for about 40 min and decorated the calendar for the coming one or two months based on their future plans. Finally, after the second session, the parents rated their children's understanding of the application again. In addition, the order of the decoration for the past and future events in each session was counterbalanced.

3 Results

3.1 Types of Parent-Child Interaction

In the analysis of application operation, we focused on the adding operation, that is, tapping the "done" button situated in the upper right corner of the display, because this was a key operation which connects the editing and adjusting operations and is the optimal operation for assessing the main operator of each group. Figure 2 shows the average percentage of the child performing the adding operation in the first and the second session in each group.

First, we calculated the percentages of the parents' and children's adding operations in each pair based on the application logs. Also, from the recorded video, it was established whether it was the parent or the child that performed the adding operation. The results showed that the children of two pairs, 4 and 6 in Table 1, performed the adding operation more than 50% of the time in the first and second sessions. These pairs were labeled *the child-led operation group*. The children of three pairs, 2, 9, and 11 in Table 1, performed the operation less than 50% of the time in the first session and more than 50% of the time in the second session. These pairs were labeled *the leader-led operation group*. Finally, the children in the remaining pairs did the operation less than 50% of the time in both sessions. These pairs were labeled *the parent-led operation group*.

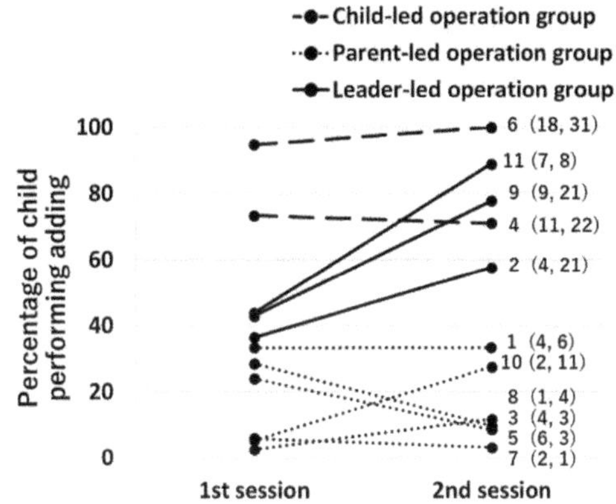

Fig. 2. Average percentage of the child's adding operation in the first and the second sessions. The numbers next to the lines show the number of the pair shown in Table 1. The numbers in the parentheses show the number of times the children performed the adding operation in the first session (left) and the second session (right).

Figure 3a shows the average rating for the children's understanding of the application in each group after the first and the second sessions. The children in the child-led and leader-led operation groups had a higher rating than the children in the parent-led operation group. Also, Fig. 3b shows the average duration that the children used the application during the break time, between the end of the first session and the beginning of the second session, and after the end of the second session.

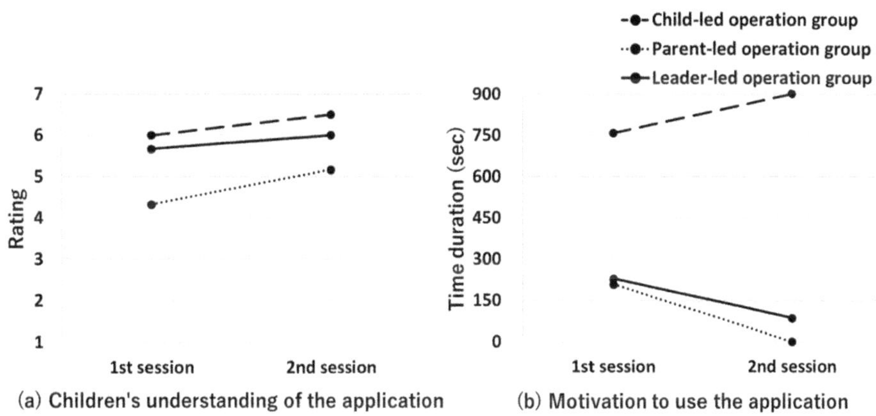

Fig. 3. (a) Average ratings for the children's understanding of the application in each group after the first and the second sessions, and (b) average duration that the children used the application during the break time, between the end of the first session and the beginning of the second session, and after the end of the second session.

In the field of education, time taken by students to perform a given task during their free time is regarded as an indicator of their motivation to learn [11]. In the child-led operation group, the amount of time the children used the application during the break time increased from the first to the second session. Moreover, there was a tendency for the children in the leader-led and parent-led operation groups not to use the application in their free time, indicating that their motivation to use the application was lower.

Since the sample size was too small to conduct statistical analyses and compare the performances among the groups, we interpreted these quantitative data together with the qualitative data based on the conversations held during the sessions in the following sections.

3.2 Conversation Analysis

Child-led Operation Group. The following case example indicates a typical utterance pattern in the child-led operation group in the first session. In this case, the parent and the child of pair 6 tried to incorporate the typed words, *opening ceremony* (of the school term), into the calendar for their past events. P and C indicate the parent and the child respectively. The number indicates the pair number shown in Table 1. Underlining indicates their body movements.

In line 1, C6 typed the words, *opening ceremony*, using the keyboard. At this point, the typed words were presented on the calendar. In line 8, P6 suggested that C6 should move the typed words, *opening ceremony*, away to somewhere, deducing that the typed words could be adjusted at this point. However, the typed words did not move, and the keyboard opened unintendedly because the "done" button had not been tapped previously to adjust the words. In line 10, P6 deduced again that the cross mark, the enter key, on the keyboard should be tapped to adjust the typed words, telling C6 to tap the cross mark. However, right after that, in line 12, P6 realized that the "done" button should be tapped before moving the words and told C6 to tap the "done" button. After that, they successfully adjusted the typed words and decorated the calendar.

In the first session, the parents in the child-led operation group tended to orally support their children's operations based on their prior knowledge and deductions, and the children actively used the application. In the second session, the interaction pattern between parents and children was the same as in the first session. However, in contrast to the first session, most of the utterances in the second session were not related to the operational methods, but they were related to the past or future events.

1	C6:	Opening ceremony. C6 typed the words *opening ceremony* using the keyboard.
2	P6:	Do you know how to write *opening ceremony* in Chinese characters?
3	C6:	I don't know.
4	P6:	If you don't, it is all right with Japanese syllabary characters. P6 pointed to a certain button to let C6 close the keyboard.
5	C6:	Right here? C6 pointed the button.
6	P6:	Yes.
7	C6:	Yes. C6 tapped the button.
8	P6:	Then, it seems like a good idea to move this away. C6 touched the typed word, and the keyboard opened.
9	C6:	Oh, what? *Opening ceremony* is not written correctly. P6 tapped the display and opened the keyboard.
10	P6:	That's right. (Tap) The cross mark. C6 tapped the cross mark, the enter key, on the keyboard. P6 tapped the cross mark again and the keyboard close button on the keyboard.
11	P6:	This should be fine. C6 tapped the display.
12	P6:	Oh, and done. P6 pointed to the "done" button to let C6 tap it.
13	C6:	What are you doing?
14	P6:	Push the cross mark. Roll this (keyboard) down. And, push "done". C6 tapped the enter key and the keyboard close button on the keyboard.
15	P6:	Now, does this move? Yes. C6 tapped the "done" button and moved the typed words.

Parent-led Operation Group. The following case example shows a typical utterance pattern in the parent-led operation group in the first session. In this case, the parent and the child of pair 7 drew words on the calendar for their past event.

In line 1, P7 opened the drawing window and asked C7 to choose what to draw and its color. In line 2, C2 decided to write her name. In line 3, P2 accepted the idea and chose the color for C2. In line 4, C7 wrote her name on the calendar. In line 7, P7 had a little difficulty with the adding operation. However, P7 randomly tapped the "done" button and managed to add the name to the calendar.

In the first session, the parents in the parent-led operation group tended to use the application most of the time and only let the children participate in operations intermittently. In the second session, the interaction pattern between parents and children was same as in the first session. Surprisingly, the children in five pairs out of six pairs of this group, pairs 1, 3, 7, 8, and 10 in Table 1, stopped working on the application in the middle of the second session.

1	P7:	Do you want to draw something? No? What color?
		P7 opened the drawing window.
2	C7:	Well, I am going to write <her name>.
3	P7:	Yes, write your name.
		P7 chose color blue to draw.
4	C7:	<her name>.
		C7 drew her name on the calendar.
5	P7:	Yes.
6	C7:	Look. I am not finished writing. I will write. Here we go.
7	P7:	Wait a minute. Done.
		P7 took a look at the explanation manual, tapped a few places, and tapped the "done" button.

Leader-led Operation Group. The following case example shows a case where the parent and the child of pair 2 tried to decorate the calendar with an emoji related to a barbecue for their future event.

In line 1, P2 said that he wanted to decorate the calendar with an emoji. In line 2, C2 asked how to do it. From lines 3 to 8, P2 actively searched for a way to decorate the calendar with an emoji by himself even tearing C2's hand away. In line 8, P2 realized how to add an emoji on the calendar. Once P2 understood the operational method, P2 let C2 do the operations, supporting C2's operations orally. In line 11, C2 looked for and found an appropriate emoji from the aligned the emojis displayed on the keyboard. In line 12, P2 told C2 to tap the "done" button to add the emoji to the calendar.

In the first session, the parents in the leader-led operation group tended to actively operate the application until they completely understood the operational methods. On the other hand, in the second session, the parents tended to transfer the operation to the children. Also, their conversations became more about the decorations than the operational methods. The parents and the children often gave their opinions about the decorations or insisted on taking over the operation from the other person.

1	P2:	I want to put the face (of emoji).
2	C2:	How can we do it?
3	P2:	I don't know how. Wait, I got it. I got it. Tap "cancel". There might be something. No, this was returned. <u>P2 canceled the earlier operation and started to tap the represented icons and words one-by-one from the upper left corner of the display. C2 tried to pull the tablet close to herself.</u>
4	P2:	Wait. Let it go. <u>P2 shook off C2's hand from the tablet.</u>
5	C2:	It should be somewhere around here. <u>C2 pointed to a certain area on the display.</u>
6	P2:	OK. Let me tap around here. No. This will take a picture. What's this? Text size. Ah, we can enter words from here. We can write barbecue here. <u>P2 tapped icons and words one-by-one from the upper left corner and opened the text editing window.</u>
7	C2:	I can't write. <u>C2 moved her pointing finger on the display.</u>
8	P2:	Wait. Probably, the typed words can be moved later. Wait. We might be able to enter words by tapping this. <u>P2 continued to shake C2's hand off from the tablet, typed barbecue on the keyboard, and handed the tablet back to C2.</u>
9	C2:	Barbecue?
10	P2:	That could be something. <u>P2 peered into the display.</u>
11	C2:	I want to do it. Barbecue. There are so many meats. There it is. <u>C2 scrolled the aligned the emojis on the keyboard, and tapped one of the emojis.</u>
12	P2:	And then, tap done, done. <u>C2 tapped the "done" button.</u>

4 Discussion

4.1 Parental Support and Child Learning Performance

In the child-led operation group, the parents tended to orally support their children's operations based on their prior knowledge and deductions. The results of the conversation analysis corresponded with those of the quantitative analysis, showing that the children had no problem in using the application. The parents accepted the children's autonomy and guided them to achieve their goals. Such children's learning through their independent activities is an effective learning strategy known as active learning which enhances learning motivation [12]. The children's active learning is considered to lead to their success in using the application and their higher motivation to use it.

In the parent-led group, the parents used the application most of the time, and the children were highly controlled and limited in using the application. This was assumed to happen because of the children's ages. All the four children under seven years old belonged to this group, and therefore, most of the parents might have thought that using the application was too difficult for their children. However, as in the quantitative analysis, the parents rated their children's understanding higher than the median rating, the middle number of the scale. This might happen because they had lower standards for their children according to their ages. Moreover, almost all the children in this group quit using the application in the middle of the second session. Such suspension of activity occurred only in this group. The previous study about motivation shows that prior achievement influences subsequent motivation [13]. The children in this group had only a few opportunities to use the application and were assumed to have had less success and consequently a less satisfying experience, which reduced their motivation to use it.

In the leader-led group, in the first session, the parents tended to use the application actively to answer the questions which they raised by themselves. In contrast, the children tended only to observe their parents' operations. In the second session, the parents tended to transfer the operation to the children. Also, their conversations became more about the decorations than the operational methods of the application. The results of the conversation analysis corresponded with those of the quantitative analysis, indicating that the children had no problem in using the application. Furthermore, in the second session, the parents and the children often gave their ideas or opinions about the decorations or insisted on taking over from the other person. These ideas, opinions or insistences were treated equally between the parents and the children. Such relationship appeared to be that of teammates [7] or project collaborators [8] in which one engages in the other's activities as a participant. There could be a possibility that the children might build a strong partnership with their parents to use the application. As a result, such a strong partnership might lower their motivation to use application independently.

4.2 Applications to Tutoring Systems

The results of this study can contribute to the development of tutoring systems in terms of operational support for children's learning and giving assistance to parents. Regarding operational support for children's learning, children need to acquire skills to use digital devices in order to learn with the devices. Therefore, tutoring systems which provide operational support are considered to be effective. Since parents' oral support enhanced children's use of a new application and motivation to use it in this study, tutoring systems which provide operational support that take children's autonomous operations into account are assumed to be efficient. Such operational support might be provided by means of simple visual texts or auditory sounds. Additionally, because younger children easily became bored when using a new application in this study, we considered that, for lower-aged children such as preschoolers, tutoring systems should give a higher priority to enhancing motivation to learn with digital devices than to teach learning content. Providing visual or auditory rewards even for minor successes could enhance their motivation to learn.

Moreover, in regard to giving assistance to parents to enhance their parental support, by providing feedback, such as the tutor programs used by children or the duration and frequency of children using certain functions or operations, to parents from tutoring systems, parents would have a better understanding of their children's skill level in using digital devices as well as their ability to learn content. Therefore, they should be able to give more appropriate support to their children. The parental mediation of children's learning support might improve their learning with tutoring systems.

Acknowledgment. We would like to thank Shogo Imamura for his contribution in planning and holding the workshop in this study.

References

1. Cabinet Office: Survey report on actual situation of early age children's internet use in 2017. Government of Japan (2018, in Japanese)
2. Rideout, V.: The Common Sense Census: Media Use by Kids Age Zero to Eight. Common Sense Media, San Francisco (2017)
3. Joanna, M., Plowman, L., Christine, S.: Just picking it up? young children learning with technology at home. Camb. J. Educ. **38**(3), 303–319 (2008)
4. Vandewater, E.A., Rideout, V.J., Wartella, E.A., Huang, X., Lee, J.H., Shim, M.S.: Digital childhood: electronic media and technology use among infants, toddlers, and preschoolers. Pediatrics **119**(5), e1006–e1015 (2007)
5. Lauricella, A.R., Wartella, E., Rideout, V.J.: Young children's screen time: the complex role of parent and child factors. J. Appl. Dev. Psychol. **36**, 11–17 (2015)
6. Flynn, R.M., Richert, R.A.: Parents support preschoolers' use of a novel interactive device. Infant Child Dev. **24**(6), 624–642 (2015)
7. Hiniker, A., Lee, B., Kientz, J.A., Radesky, J.S.: Let's play! digital and analog play patterns between preschoolers and parents. In Proceedings of the SIGCHI Conference on Human Factors in Computing Systems, pp. 1–13. ACM Press, NY (2018)
8. Barron, B., Kennedy-Martin, C., Takeuchi, L., Fithian, R.: Parents as learning partners in the development of technological fluency. Int. J. Learn. Media **1**(2), 55–77 (2009)
9. Takeuchi, L., Stevens, R.: The New Coviewing: Designing for Learning Through Joint Media Engagement. Joan Ganz Cooney Center at Sesame Workshop, NY (2011)
10. Tojo, N., Ishizaki, H., Nagai, Y., Niida, N.: Tool for enhancing family communication through planning, sharing experiences, and retrospection. In the 2nd International Conference on Computer-Human Interaction Research and Applications, pp. 34–44. SciTePress, Setúbal (2018)
11. Harackiewicz, J.M., Elliot, A.J.: Achievement goals and intrinsic motivation. J. Pers. Soc. Psychol. **65**(5), 904–915 (1993)
12. Elliot, A.J., Harackiewicz, J.M.: Approach and avoidance achievement goals and intrinsic motivation: a mediational analysis. J. Pers. Soc. Psychol. **70**(3), 461–475 (1996)
13. Martin, A.J., Way, J., Bobis, J., Anderson, J.: Exploring the ups and downs of mathematics engagement in the middle years of school. J. Early Adolesc. **35**(2), 199–244 (2014)

PKULAE: A Learning Attitude Evaluation Method Based on Learning Behavior

Deqi Li, Zhengzhou Zhu(✉), Youming Zhang, and Zhonghai Wu

School of Software and Microelectronics, Peking University, Beijing, People's Republic of China
zhuzz@pku.edu.cn

Abstract. Learning attitude is an important factor related to students' academic achievement. The existing E-learning systems has paid little attention to the students' learning attitude, which lacks the ability of supervision for students' learning behavior. This paper designs a learning attitude evaluation method PKULAE based on 17 kinds of learning behaviors. This method analyses the students' learning behavior from both global and local aspects. It judges the student's learning attitude from global and local aspects by regression analysis and GBDT regression tree constructed by the students' behavior characteristic matrix. Finally, it determines the students' learning attitude in a certain period by voting. The paper classifies learning attitudes into positive attitudes and negative attitudes. We did experiments based on the online behavior data of 125 students in Peking University. The result shows that PKULAE method has been significantly improved compared with GBDT and RF. When the recall and the TNR are no less than GBDT and RF, the accuracy is 0.778, which is at least 0.07 higher than GBDT and RF.

Keywords: Learning attitude · Learning behavior analysis · Regression analysis · Software engineering education · GBDT regression tree

1 Introduction

Learning attitude refers to a relatively stable psychological tendency formed by students in their own learning process, including cognitive, emotional and behavioral factors [1].

In traditional teaching environment, the evaluation of students' learning attitude mainly depends on teachers' personal observation, which makes it difficult to give timely attention to and rectify all abnormal students who don't pay their attention to learning. Now a large amount of learning behavior data have been generated in e-learning platforms. Based on these data, we designs a learning attitude judgment method, which can effectively analyze students' learning attitude, make up for the function of the current e-learning platform, and provide a better learning environment for students' learning.

A. Coy et al. (Eds.): ITS 2019, LNCS 11528, pp. 156–162, 2019.
https://doi.org/10.1007/978-3-030-22244-4_19

2 Related Works

Learning attitude is an important factor affecting students' academic achievement. It has a very close relationship with students' learning behavior. It also affects students' learning efficiency and learning effect [2]. It is the main factor affecting students' learning behavior [3]. According to the 2007 CRUMP survey sponsored by the Ministry of Education and Science of Japan, students' learning situation includes attitude and behavior. Attitude determines the content and manner of behavior, while behavior reacts on attitude [4]. Zhang Ping and others believe that students' learning attitude determines their learning direction and learning attitude can reflect students' learning style and behavior to a certain extent [5].

At present, the main research on learning attitude is qualitative analysis, which judges the students' learning attitude by artificial way. The most common and direct evaluation index of learning attitude is academic achievement [6]. However, the purpose of studying learning attitude is to improve academic achievement. Therefore, before achieving academic achievement, we should evaluate learning attitude from more angles. In addition to using academic achievement directly, other researchers pay more attention to evaluating students' learning attitudes through some specific questions, mainly through questionnaires. Hwang [7] evaluates students' e-learning attitude through questionnaires. Literature [8] Evaluation of nursing students' e-learning attitude through interviews and case analysis. Although many researchers have conducted surveys and assessments on the evaluation of learning attitudes, they are unable to get rid of the two problems of personal subjective factors and quantification, and lack a set of quantitative and objective means to evaluate students' learning attitudes, either from the perspective of achievement or from the perspective of questionnaires.

3 Design of PKULAE Method

This paper defines online operation and head movement as learning behavior, which occurs in the process of watching videos. There are 17 kinds of behaviors in total. The learning behavior data is divided into two parts: the global and the local. Local learning behavior reflects learning attitude when learning specific knowledge points. Overall learning behavior reflects students' learning attitude towards curriculum, specialty and even all e-learning content in the learning process. Combining the two learning behavior characteristics, this paper designs a learning attitude evaluation method named Peking University Learning Attitude Evaluation (PKULAE).

3.1 Analysis of Local Learning Attitudes

Local learning attitude relies on the data of learning behavior and watching time. The learning behavior data DATALB is used as the training data of local learning attitude, the time data is used as the weight data, and the regression results are further processed as the weight when the local learning attitude is finally judged.

Considering that students' learning behavior has a certain time series, assuming that x is the learning behavior of students at the current moment and y is the learning behavior of students at the moment before x, then (y, x) is a learning behavior sequence pair of students. Therefore, the probability of the student's current learning behavior x occurring is shown in formula (1).

$$P(x|y) = \frac{P(xy)}{P(y)} = \frac{\left[\frac{n(xy)}{n2}\right]}{\left[\frac{n(y)}{n1}\right]} \tag{1}$$

Among them, $P(xy)$ is the probability of learning behavior sequence pairs (y, x) appearing in the total learning behavior sequence pairs, $P(y)$ represents the probability of behavior y appearing in the total learning behavior pairs, and $n(xy)$, $n(y)$ is the number of occurrences of learning behavior sequence pairs (y, x) and learning behavior y in the current data. $n2$ is the number of all sequence pairs and $n1$ is the number of all behaviors.

Assuming that every student in the system has m kinds of learning behavior, each student has an $m*m$ behavior characteristic matrix, as shown in Fig. 1(a), each element in the matrix is a learning behavior probability. Since there is no sequence relationship between the probabilities in the feature matrix, it can be transformed into an m^2-dimensional feature vector, as shown in Fig. 1(b). In order to reduce over-fitting, principal component analysis is used to reduce the dimension of the vector. With 97% feature information preserved, the m^2 dimension feature vector can be reduced to the l dimension feature vector. As shown in Fig. 1(c), it is the input data for constructing GBDT regression tree.

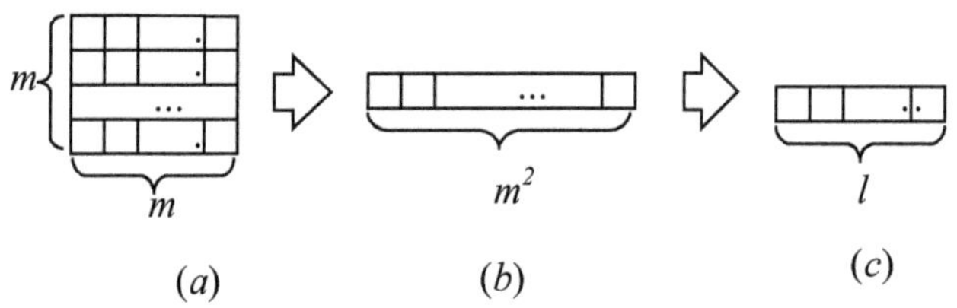

Fig. 1. Schematic diagram of training data preprocessing

Each student can produce an l-dimension eigenvector. The vector is input into the regression tree(GBDT), and the preliminary regression result rst_1 can be obtained.

Assuming that the watching time is t, the lower limit of watching time is τ_0. If students' one-cycle watching time is less than τ_0, they think that the students are not fully conscientious. Therefore, according to the data of the students' watching time, the weight of the regression results w_1 can be determined by formula (2).

$$w_1 = \frac{n_{t \geq \tau_0}}{n_{total}} \tag{2}$$

Among them, $n_{t \geq \tau_0}$ denotes the number of cycles that students learn longer than the lower limit of learning time, and n_{total} denotes the number of cycles of curriculum learning.

According to formula (3), the regression result rst_1 is normalized to the maximum value, and all local learning attitude analysis results of the data set to be tested are rst_{local} $(0 < rst_{local} < 1)$.

$$rst_{local} = rst_1 * w_1 / \max\left(rst_1' * w_1'\right) \tag{3}$$

Among them, $\max\left(rst_1' * w_1'\right)$ represents the maximum value in the current detection data set. rst_1' represents the set of analysis results of all students' partial learning attitude and w_1' represents the weight set of analysis results of all students' partial learning attitude.

3.2 Analysis of Overall Learning Attitude

The analysis of holistic learning attitude focuses on the distribution of learning content and head posture during the global e-learning process. Normally, students have a similar distribution. Students with negative learning attitudes tend to be more perfunctory in the learning process, such as concentrating on one or two courseware, which will be hard to fit the distribution. Statistics of each student's e-learning situation, get the data set DATA_{LTD}, DATA_{LTD} in each student's e-learning time distribution, according to the cycle of statistics of each student's e-learning data frequency, according to the order from big to small, sorting each student's data frequency, curve regression of data, get regression curve, using R square as regression evaluation index, remember.

DATA_{LFP} is acquired in the process of students' e-learning, and the key points of face are extracted by OpenPose. According to the key points of face, eyebrow spacing, lip-nose ratio and face length-width ratio are stored for each user. By analyzing these data, students can be divided into face or not face the screen. By dividing the number of pictures that face the screen in a period of time by the number of all pictures in that period, we can determine the proportion of time that students watch the screen w_2. Finally, the final overall learning attitude rst_{global} is obtained by formula (4).

$$rst_{whole} = rst_2 * w_2 / \max\left(rst_2' * w_2'\right) \tag{4}$$

Among them, $\max\left(rst_2' * w_2'\right)$ represents the maximum value in the current detection data set and plays the role of maximum normalization. rst_2' denotes the set of the results of the analysis of the overall learning attitude of all students, and w_2' denotes the weight set of the results of the analysis of the overall learning attitude of all students.

3.3 Generate Final Learning Attitude

In this paper, GBDT and curve regression are used to replace the original weak classifier in bagging algorithm. The results of rst_{local} and rst_{global} from the two classifiers are integrated by voting after dichotomy to generate the final learning attitude.

We set δ as the threshold, which is between 0 and 1. Only both of rst_{local} and rst_{global} 's scores reach above δ, can the student be considered as positive learning attitude during this period as shown in the Table 1.

Table 1. Analysis of binary learning attitudes.

rst_{global}	rst_{local}	
	[0, δ)	[δ, 1]
[0, δ)	Negative	Negative
[δ, 1]	Negative	Positive

4 Experiments and Verification

We build an e-learning platform based on Moodle. The learning content is 116 teaching videos of Software Engineering including 53 lectures. Each video is about 20 min. The125 participants were undergraduates majoring in software engineering of PKU. According to the performance management method promulgated by the Peking University, above 80 points are considered good, and below 80 points are not very good or unqualified. As a result, only both of two regression result score reach above 0.8, we consider the attitude is positive. When the experiment is carried out, a week is taken as a learning cycle, and the length of study in each cycle is defined as one hour. Since there are 17 learning behaviors defined, a 17*17 behavior feature moment is obtained in the local feature analysis. Then it's reformed into the shape of input data.

In the experiment, the data set was divided by the method of set-aside. 36 students were randomly selected from 125 students at a time as test samples and 89 students as training samples. According to the practical application purpose of the analysis results of learning attitude, the accuracy and recall are used as the evaluation criteria. *Acc* is an evaluation criterion to measure whether it can effectively analyze students' two states. The recall R and the *TNR* (true negative rate) can reflect which learning attitude the algorithm pays more attention to, and can analyze student data pertinently.

A total of 4 rounds of experiments were carried out, 36 students were randomly selected for each round to test. RF (Random Forests) algorithm and GBDT algorithm were compared, and 80/0.8 was used as the threshold to classify attitudes. The experimental results are shown in Fig. 2.

According to the experimental results, the average evaluation results of this method, GBDT method and RF method in Table 2 are calculated. From the table, we can see that the overall accuracy of this method is better than the results of the analysis using GBDT or RF alone. When analyzing students with positive learning attitudes, both the method and GBDT have produced good experimental results. The recall reaches 0.804, while the RF analysis results only have 0.500 recall. For students with

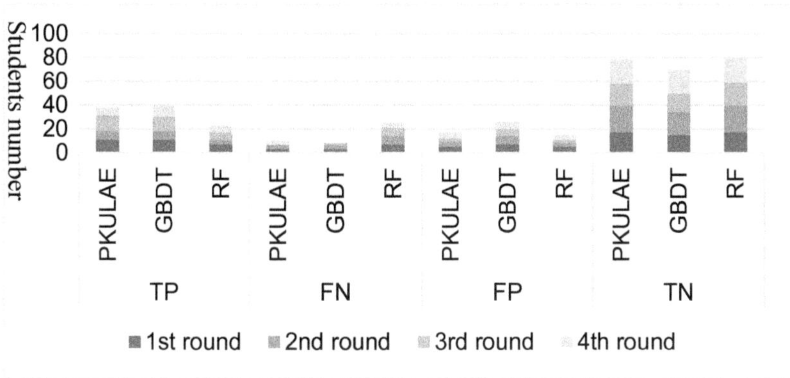

Fig. 2. The TP, FN, FP, TN frequency of PKULAE, GBDT and RF

negative learning attitude, the results of PKULAE and RF are better than those of GBDT. The TNR is 0.761 and 0.784 respectively. From the data comparison, we can find that this method can achieve better results than the popular GBDT algorithm and RF algorithm for students with positive or negative learning attitudes.

Table 2. Comparisons of experimental results of three methods.

	Acc	R	TNR
PKULAE	0.778	0.804	0.761
GBDT	0.708	0.804	0.648
RF	0.673	0.500	0.784

5 Conclusion

This paper proposed a learning attitude evaluation method PKULAE. In the process of local learning attitude analysis, the feature matrix is constructed using the students' operational data, and the input data of GBDT algorithm is obtained by dimensionality reduction using principal component analysis. Then the regression tree is constructed. Finally, the results of local learning attitude analysis are obtained by using the time-length weight and maximum normalization method. The results of overall learning attitude analysis were obtained by curve regression method, and then the maximal normalization of face-to-face time ratio was used as weight to obtain the overall learning attitude results. The voting method is used to vote on the two results to obtain the final students' learning attitude.

Acknowledgments. This paper was supported by National Key Research and Development Program of China (Grant No. 2017YFB1402400), Ministry of Education "Tiancheng Huizhi" Innovation Promotes Education Fund (Grant No. 2018B01004), National Natural Science Foundation of China (Grant No. 61402020), and CERNET Innovation Project (Grant No. NGII20170501).

References

1. Liu, H., Zhao, W., Wang, L.: An empirical study on effects of teacher expectation on students' learning attitudes in blended learning environment. J. Distance Educ. **32**(1), 50–51 (2014)
2. Li, D., Zheng, W., Zhu, J.: A study on the learning attitudes of postgraduates and their relations with school factors, take Shenzhen S University as an example. High. Educ. Explor. **33**(7), 73–79 (2017)
3. Li, X., Guo, J.: Study on the correlativity of students' learning attitude and behavior. Stud. Psychol. Behav. **3**(4), 265–267 (2005)
4. Dou, X.H., Motohisa, K., Mio, H.: A study on learning behavior and attitude of the contemporary Japanese College students: an analysis of the national survey of Japanese undergraduates in 2007. Fudan Educ. Forum **9**(5), 79–85 (2011)
5. Zhang, P., et al.: Reforming the teaching model and promoting the positive development of students' attitudes towards physics learning. China Univ. Teach. **36**(2), 37–40 (2014)
6. Williams, E.: Student attitudes towards approaches to learning and assessment. Assess. Eval. High. Educ. **17**(1), 45–58 (1992)
7. Hwang, G.J., Chang, H.F.: A formative assessment-based mobile learning approach to improving the learning attitudes and achievements of students. Comput. Educ. **56**(4), 1023–1031 (2011)
8. Tenison, E., Touger-Decker, R.: Impact of e-Learning or blended learning versus face-to-face learning in regard to physical examination skills, knowledge, and attitudes among health professions students. Top. Clin. Nutr. **33**(3), 259–270 (2018)

Predicting MOOCs Dropout Using Only Two Easily Obtainable Features from the First Week's Activities

Ahmed Alamri[1], Mohammad Alshehri[1], Alexandra Cristea[1(✉)], Filipe D. Pereira[2], Elaine Oliveira[2], Lei Shi[3], and Craig Stewart[4]

[1] Department of Computer Science, Durham University, Durham, UK
alexandra.i.cristea@durham.ac.uk
[2] Institute of Computing, Federal University of Roraima, Boa Vista, Brazil
[3] Centre for Educational Development, University of Liverpool, Liverpool, UK
[4] School of Computing Electronics and Mathematics, Coventry University, Coventry, UK

Abstract. While Massive Open Online Course (MOOCs) platforms provide knowledge in a new and unique way, the very high number of dropouts is a significant drawback. Several features are considered to contribute towards learner attrition or lack of interest, which may lead to disengagement or total dropout. The jury is still out on which factors are the most appropriate predictors. However, the literature agrees that early prediction is vital to allow for a timely intervention. Whilst feature-rich predictors may have the best chance for high accuracy, they may be unwieldy. This study aims to *predict learner dropout early-on, from the first week*, by comparing several machine-learning approaches, including Random Forest, Adaptive Boost, XGBoost and GradientBoost Classifiers. The results show *promising accuracies (82%–94%) using as little as 2 features*. We show that the accuracies obtained outperform state of the art approaches, even when the latter deploy several features.

Keywords: Educational data mining · Learning analytics · Dropout prediction · Machine learning · MOOCs

1 Introduction

A key concept of MOOCs is to provide open access courses via the Internet that can scale to any number of enrolled students [1]. This vast potential has provided learning opportunities for millions of learners across the world [2]. This potential has engendered the creation of many MOOC providers (such as FutureLearn, Coursera, edX and Udacity)[1], all of which aim to deliver well-designed courses to a mass audience. MOOCs provide many valuable educational resources to learners, who can connect and collaborate with each-other through discussion forums [3]. Despite all their benefits, the rate of non-completion is still over 90% for most MOOCs [4]. Research is still

[1] https://www.mooclab.club/resources/mooclab-report-the-global-mooc-landscape-2017.214/

A. Coy et al. (Eds.): ITS 2019, LNCS 11528, pp. 163–173, 2019.
https://doi.org/10.1007/978-3-030-22244-4_20

undergoing on whether the low rate of completers indicates a partial failure of MOOCs, or whether the diversity of MOOCs learners may lead to this phenomenon [2]. In the meantime, this problem has attracted more attention from both MOOC providers and researchers, whose goal is to investigate methods for *increasing completion rates*. This starts by determining the *indicators of student dropout*. Previous research has proposed several indicators. Ideally, the earlier the indicator can be employed the sooner the intervention can be planned [5]. Often, combining several indicators can raise the precision and recall of the prediction [6]; however, such data may not always be available. For example, a linguistic analysis of discussion forums showed that they contain valuable indicators for predicting non-completing students [7]. Nevertheless, these features are not applicable to the majority of the student population, as only five to ten percent of the students post comments in MOOC discussion forums [8]. In this paper, we present *a first of its kind research into a novel, light-weight approach based on tracking two* (*accesses to the content pages* and *time spent per access*) *early, fine grained learner activities to predict student non-completion*. Specifically, the machine learning algorithms take into account the first week of student data and thus are able to 'notice' changes in student behaviour over time. It is noteworthy that we apply this analysis on a MOOC platform firmly rooted in pedagogical principles, which has seen comparatively less investigation, namely FutureLearn (www.futurelearn.com). Moreover, we apply our method on a large-scale dataset, which records behaviour of learners in very different courses in terms of disciplines. Thus, the original research question this study attempts to address is:

RQ. *Can MOOC dropout be predicted within the first week of a course, based on the learner's number of accesses and time spent per access?*

2 Related Research

MOOCs' widespread adoption during their short history, has offered the opportunity for researchers and scientists to study them; with specific focus given to their low rate of completion. This has resulted in the creation of several predictive models that determine student success, with a substantial rise in the literature since 2014 [9].

Predicting students' likelihood to complete (or not to complete) a MOOC course, especially from very early weeks, has been one of the hottest research topics in the area of learning analytics. Kloft et al. [2] used the weekly history of a 12-week-long psychology MOOC course to notice changes in student behaviours over time, proposing a machine learning framework for prediction of dropout and achieving an increase by 15% in prediction accuracy (up to 70%–85% for some weeks) when compared to baseline methods. However, the model proposed didn't perform correctly during the early weeks of the course. Hong et al. [10] proposed a technique to predict dropouts using learning activity information of learners via applying a two-layer cascading classifier; three different machine learning classifiers - Random Forest (RF), Support Vector Machine (SVM) and Multinomial Logistic Regression (MLR). This study achieved an average of 92% precision and 88% accuracy predicting student dropout. Xing et al. [11], considered active students who were struggling in forums, by designing a prioritising at-risk student temporal modelling approach. This aims to provide systematic insight for instructors to target those learners who are most in need of intervention. Their study illustrates the effectiveness of an ensemble stacking generalisation approach to build more robust and accurate prediction models. As most research on MOOC dropout prediction has measured test accuracy on the same course used for training, this can lead to overly optimistic accuracy estimates. Halawa et al. [12] designed a dropout predictor using student activity features for timely intervention delivery. The predictor scans student activities for signs of lack of ability or interest, which may lead to long periods of absence or complete dropout. They identified 40%–50% of dropouts while learners were still active. However, the results provided often failed to specify the precision and recall, or, if they did, they were not detailed at the level of a class (such as for completers and non-completers, separately), but averaged. This is an issue, as it introduces a potential bias, which we further discuss later in this paper.

Additionally, the data is seldom balanced between the classes. This is yet another problem, specifically for MOOCs, where the data distribution between the classes is so skewed (with around 90% of the students belonging to the non-completers class, and only 10% completers). In combination with the averaging of the results, this could lead to over optimistic results. Hence in this paper, we report the results in detail at class level, as well as balancing the data across the classes.

In terms of best performing learning algorithms, the use of random forest (RF) (e.g., [13–16]) has appeared in the literature among the most frequently used approaches for the student classification tasks. Additionally, Ensemble Methods, such as boosting, error-correcting have been shown to often perform better than single classifiers, such as SVM, KNN and Logistic Regression [17, 18]. In this sense, and to support our early prediction, low feature number approach, we applied the following state-of-the-art classification algorithms to build our model, moving them to the education domain: RF, GradientBoost, AdaBoost and XGBoost. Further improving on the algorithms may render higher accuracy, but is beyond of the scope of this paper.

There have been other studies that have proposed using several machine learning techniques at the same time, to build their prediction models. One study [19] used four different machine learning techniques, including RF, GBM, k-NN and LR, to predict which students are going to get a certificate. However, they used a total of eleven independent variables to build the model and predict the dependent variable – the acquisition of a certificate (true or false); whereas our model uses only two independent variables (the number of accesses and the time spent on a page). Additionally, their results indicated that most learners who dropped out were likely to do so during the first weeks. This supports our assumption that early prediction is possible and can be accurate. Importantly, unlike our approach of using only two independent variables (features/attributes), most prior research used many. For example, [2] employed nineteen features, including those that capture the activity level of learners and technical features. Promisingly our model, despite using only two features from only the first week of each course, can also achieve a 'good enough' performance, as shall be further shown.

3 Methodology

3.1 Data Preparation

This study has analysed data extracted from 21 runs of 5 FutureLearn-delivered courses from The University of Warwick between 2013 and 2017. The number of accesses and the time spent have been computed for each student. The courses analysed can be roughly classified into 4 main themes: literature (Shakespeare and His World); Psychology (The Mind is Flat) and (Babies in Mind); Computer Science (Big Data) and Business (Supply Chains). Runs represent the number of repeated delivery for each of the five courses. The number of runs for each course is (5, 6, 6, 3 and 2, respectively) whereas the set number of weeks required for studying each course is (10, 6, 4, 9 and 6). In total, they involve the activities of 110,204 learners, who accessed 2,269,805 materials, with an average of around 21 materials accessed per student.

Some courses offer quizzes every week, on subjects of different nature and/or difficulty level, whereas others skip some of the weeks. Due to all the above variations between the courses, we have considered it best to analyse each courses independently, merging only the data from different runs of each course. The latter was made possible, as all courses had runs (within that course) of identical length and similar structure.

In order to determine if there is a normal distribution of variables in each group (completers and non-completers), the Shapiro–Wilk test was used. On determining that distribution was non-parametric, the Wilcoxon signed-rank test was applied, to determine if there is a significant difference between completers and non-completers.

In order to prepare and analyse the data, we next define the employed feature extraction and selection technique, as well as the machine learning algorithms previously identified to address our research question. To begin with, the raw dataset was refined, removing all students who enrolled but never accessed any material. We dealt with those learners separately, based on even earlier parameters (such as the registration date) [20]. Subsequent to this there were 110,204 remaining learners to be studied, of which 94,112 have completed less than 80% and only 16,092 have completed 80% or more of the materials in the course. The reason of selecting 80% completion as a sufficient level of completion (as opposed to, e.g., 100% completion) is based on prior literature and our previous papers [20–22], where we consider different ways of computing completion. Moreover, the total number of those who completely accessed 100% of the steps was relatively low.

In terms of early prediction, we have opted for the first week, as this methodology is one of the most difficult and least accurate approaches when comparing with the current state of the art in the literature. Alternatively, a relative length (e.g., 1/n days of the total length of each course) could have been used. However, in practice, this tends to use later prediction data than our approach (e.g., 1/4th of a course is 1 week for Babies in Mind, but 2.5 weeks for Shakespeare and his Work).

3.2 Features Selection

Unlike the current literature, this study determined to minimise the number of indicators utilised. In order to check which indicators are more important, we use an embedded feature selection method that evaluates the importance of each feature by the time that the model is training. As we used tree-based ML algorithms, the metric to measure the importance of each feature was the Gini-index [23]. Figure 1 shows the most important features for each course.

As one of the goals of this study was to create a simple model, we focused on specific features which could be used for various MOOCs – this was done to enhance the generalisation and applicability of the findings for the providers. Therefore, we applied four features to predict the student completion, as follows. ***Number of Accesses*** represents the total number of viewed steps (articles, images, videos), whereas ***Time Spent*** represents the total time spent to complete each step. ***Correct answers*** represents the total number of correct answers and ***Wrong answers*** represents the total number of wrong answers (see Fig. 1 Gini-importance for all the five courses).

We concluded that ***Time spent***, and ***Number of access*** are the most important features, since those two features are not only easy to obtain for most courses, but also results show that ***Time spent*** in each step is playing a critical role to predict the student completion. Moreover, the number of accesses was, in general, an important feature in all the courses. Furthermore, it should be taken in consideration that some courses do not have quizzes in every week; in this case the ***Wrong answer*** and ***Correct answers*** features do not play any role to predict the student's completion in those courses (see big data course in Fig. 1(*d*)).

3.3 Building Machine Learning Models

To build our model, we employed several competing ML ensembles methods, as follows: Random Forest (RF) [27], Gradient Boosting Machine (Gradient Boosting), [24] Adaptive Boosting (AdaBoost) [25] and XGBoost [17] to proceed with exploratory analysis. Ensembles refers to those learning algorithms that fit a model via combining several simpler models and converting weak learners into strong ones [26]. In cases of binary classification (like ours), Gradient Boosting uses a single regression tree to fit on the negative gradient of the binomial deviance loss function [24]. XGBoost, a library for Gradient Boosting, contains a scalable tree boosting algorithm, which is widely used for structured or tabular data, to solve complex classification tasks [17]. Adaboost is another method, performing iterations using a base algorithm. In each interaction, Adaboost uses higher weights for samples misclassified, so that this algorithm focuses more on difficult cases [25]. Random Forest is a method that use a number of decision trees constructed using bootstraping resampling and then applying majority voting or averaging to perform the estimation [27].

After comparing the above methods based on a training and test set division of 70%/30% respectively, in order to more accurately estimate the capacity of the different methods to generalise to an independent (i.e., unknown) dataset (e.g., to an unknown run of a course), and to avoid overfitting, we have also estimated the prediction accuracy based on 10-fold cross-validation, a widely used technique to evaluate a predictive model [28]. In order to obtain confidence intervals for all the performance metrics (accuracy, precision, recall, F1-score), we have attempted to predict student completion a hundred times, by choosing testing and training sets randomly [29].

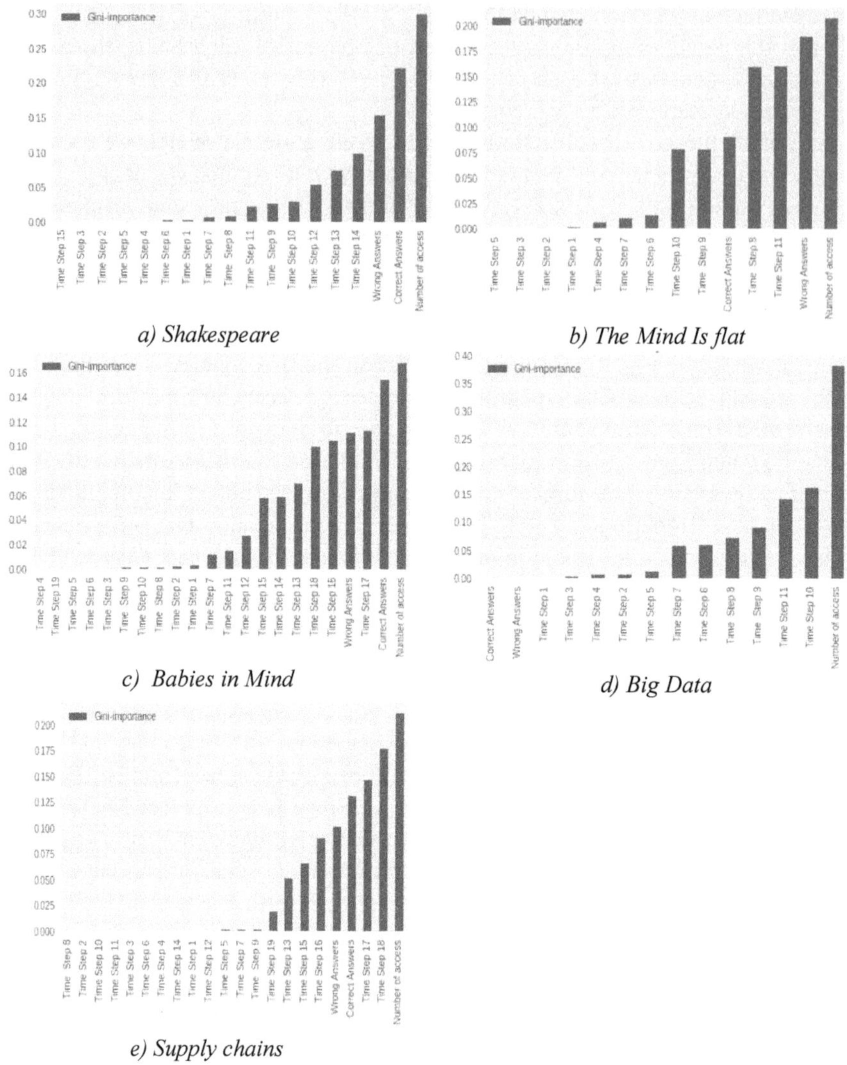

Fig. 1. Gini-index for the features in the five courses.

4 Results

This section details the results of our prediction task of using the first week to determine if the learners selected in the above section are to be completers or non-completers, based on different algorithms. Table 1 compares Random Forest (RF), Adaboost Classifier, XGBoost Classifier and GradientBoosting Classifier methods for all five courses, reporting on some of the most popular indicators of success: accuracy, precision, recall, and the latter two combination, the F1 score.

Table 1. Prediction performance for balanced data (oversampling)

	Accuracy		Precision				Recall				F1 Score			
Big data		[+ −]	0	[+ −]	1	[+ −]	0	[+ −]	1	[+ −]	0	[+ −]	1	[+ −]
Random forest	91.08	0.04	98	0.03	85	0.07	83	0.09	98	0.02	90	0.05	91	0.04
Gradient boosting	**91.43**	**0.04**	**99**	0.01	85	0.07	83	0.09	99	0.01	90	0.05	92	**0.04**
AdaBoost	91.37	0.04	99	0.01	85	0.07	82	0.08	99	0.01	90	0.05	92	0.04
XGBBoost	**91.38**	**0.05**	**99**	0.02	85	0.08	82	0.09	99	0.01	90	0.05	92	**0.05**
The mind is flat														
Random Forest	**87.65**	**0.05**	**98**	0.04	80	0.07	76	0.08	98	0.03	86	0.05	88	**0.04**
Gradient boosting	87.91	0.04	98	0.02	80	0.06	76	0.08	99	0.02	86	0.05	89	0.04
AdaBoost	**87.78**	**0.04**	**99**	0.03	80	0.07	76	0.08	99	0.02	86	0.05	89	**0.04**
XGBBoost	87.94	0.05	99	0.03	80	0.06	76	0.08	99	0.02	86	0.05	89	0.04
Babies in mind														
Random forest	82.69	0.05	96	0.04	75	0.08	67	0.14	97	0.03	79	0.06	84	0.05
Gradient boosting	**83.47**	**0.05**	**98**	0.04	75	0.08	67	0.1	98	0.03	80	0.07	85	**0.05**
AdaBoost	83.30	0.05	98	0.05	75	0.08	67	0.1	99	0.03	80	0.07	85	0.05
XGBBoost	**83.41**	**0.06**	**98**	0.04	75	0.08	67	0.11	99	0.02	80	0.08	85	**0.05**
Supply chain														
Random forest	**92.08**	**0.11**	**99**	0.06	86	0.17	85	0.22	99	0.05	91	0.13	92	**0.1**
Gradient boosting	93.40	0.1	99	0.03	88	0.18	86	0.2	99	0.03	92	0.11	93	0.1
AdaBoost	**93.11**	**0.1**	**99**	0.05	88	0.17	86	0.19	99	0.04	92	0.11	93	**0.1**
XGBBoost	93.14	0.09	99	0.03	87	0.16	86	0.19	99	0.02	92	0.11	93	0.09
Shakespeare														
Random forest	93.03	0.09	99	0.04	88	0.15	86	0.18	99	0.04	92	0.11	93	0.09
Gradient boosting	**93.26**	**0.11**	**99**	0.04	88	0.17	86	0.22	99	0.03	92	0.13	93	**0.1**
AdaBoost	93.10	0.1	99	0.06	88	0.16	86	0.19	99	0.05	92	0.11	93	0.09
XGBBoost	**93.20**	**0.09**	**99**	0.05	88	0.16	86	0.2	99	0.05	92	0.11	93	0.09

0: Non- Completer Group, 1: Completer Group, [+−]: Error of margin over 100 prediction times

In general, all algorithms achieved almost the same result, indicating that regardless of the employed model, the features selected in this study proved to be powerful in predicting completers and non-completers. Moreover, our predictive models were able to achieve high performance in each class (completers '1' and non-completers '0') as shown in Table 1. *The prediction accuracy varies between 83%–93%.* We can see that the best performing course, across all four methods applied, is the 'Shakespeare' course.

The chart below (Fig. 2) illustrates the median of the time spent by completers and non-completers on the first step of the first week across all the five courses. Results show that completers spent between 66% to 131% more time than non-completers in Big Data and Shakespeare, respectively. Supply Chain recorded the highest ratio between both groups of learners, with 601% more time spent by completers. However, the difference between the two groups was lower, i.e., 25% more for completers, for Babies in Mind.

Additionally, the Shapiro test was used to determine the normal distribution of variables in each group (completers and non-completers). The results show that the time spent is not normally distributed (p-value < 2.2e−16) in all courses. Therefore, the Wilcoxon test was used to determine if there is a significant difference between the completers and non-completers groups. The results show that two data sets are significantly different in all courses – in other words, that the completers spend not only more time on average than the non-completers, but that this difference is significant.

5 Discussion

We have selected four of the most successful methods for classification problems, applying them in the domain of learning analytics in general, and on completion prediction in particular.

Another candidate was SVM, which we did apply, but which was less successful with a linear kernel and would possibly need a non-linear kernel to improve accuracy. In terms of the variation of accuracy, precision, recall and F1 score between courses, the best performing course, ‘Shakespeare’, was the longest (10 weeks), with a relatively good amount of data available (5 runs). The worst performing course, on the other hand, ‘Babies in Mind’, was the shortest (4 weeks).

Thus, for all methods, long courses, such as ‘Shakespeare’ (spanning over 10 weeks) and ‘Big Data’ (taking 9 weeks), perform better. Moreover, it seems that the longer the course, the better the prediction, as the prediction for the 10-week course on Shakespeare outperforms the prediction of the 9-week course on Big Data across all methods consistently, for both training and test set. *Our accuracy is very high - between 82–94% across all courses. This is equivalent to the current best in breed from the literature, but utilised far fewer indicators to achieve a much earlier success.* This is due to the careful selection process of the two features, which are both generic, as well as informative. One important reason of why the two early, first week features were enough for such good prediction is the fine granularity of the mapping of these features – for each FutureLearn ‘step’ (or piece of content) we could compute both number of accesses as well as time spent. Thus, the application of the features for the first week transformed into a multitude of pseudo-features, which would explain the increased prediction power. Nevertheless, this method is widely applicable and does not detract from the generalisability of our findings.

Importantly, we have managed to predict only based on the first week of the course, how the outcome will look like. For some courses, this represents prediction based on a quarter of the course (e.g., for Babies in Mind). For others, the prediction is based on data from one tenth of the course, which is a short time to draw conclusions from.

A few further important remarks need considered. Firstly, the data pre-processing is vital: here we want to draw the attention especially to the balancing of the data. For such extremely skewed datasets as encountered when studying MOOC completion, where averages of 10% completion are the norm, prediction can ‘cheat’ easily: by, e.g., just predicting that all students fail, we would obtain a 90% completion rate! In order to avoid such blatant bias, we balance the data.

Furthermore, the way the data is reported is important. Many studies just report the average for the success measure (be it accuracy, recall, precision, F-score, etc.) over the two categories. As we can see above, the difficulty in the problem we are tackling is the

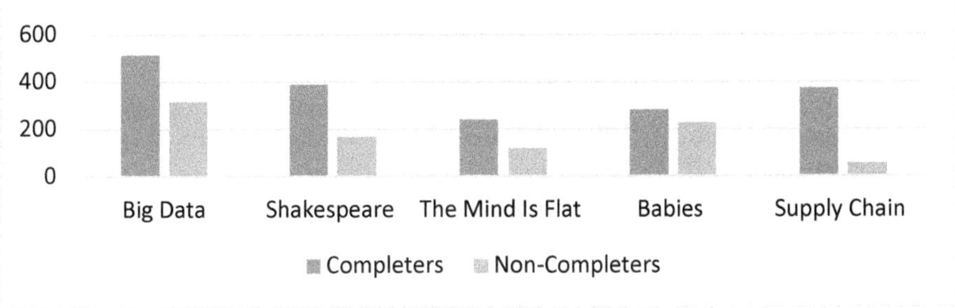

Fig. 2. Time spent on the first step of the first week (week one) by completers and non-completers.

prediction of the completers: thus, it would be easy to hide the poor prediction on this 'hard' category, by averaging the prediction across categories and students. To ensure this is not happening, we provide in this paper separate measures for each category, so the results we are reporting don't suffer from this bias.

6 Conclusion

In this paper, we have shown the results from our original study that demonstrates that we can provide reliable, very early (first week) prediction based on two easily obtainable features only, thus via a light-weight approach for prediction, which allows for easy and reliable implementation across various courses from different domains. Such an early and accurate predictive methodology does not yet exist beyond our research and as such this is the first in this class of model. We have shown that these two features can provide a 'good enough' performance, *even outperforming state of the art solutions involving several features*. The advantage of such an approach is clear: it is easier and faster to implement across various MOOC systems, and does require the existence of only a limited amount of information points. The implementation itself is light-weight, and is much more practical when considering an on-the-fly response, and has a limited cost in terms of implementation resources, and more importantly, in terms of time. The results we have obtained are based on balanced datasets, and we report success indicators across both categories, completers and non-completers. We thus avoid both bias in terms of unbalanced datasets, as well as bias based on averaging.

Acknowledgment. We would like to thank FAPEAM (Foundation for the State of Amazonas Research), through Edital 009/2017, for partially funding this research.

References

1. Ipaye, B., Ipaye, C.B.: Opportunities and challenges for open educational resources and massive open online courses: the case of Nigeria. Commonwealth of Learning. Educo-Health Project. Ilorin (2013)

2. Kloft, M., Stiehler, F., Zheng, Z., Pinkwart, N.: Predicting MOOC dropout over weeks using machine learning methods. In: Proceedings of the EMNLP 2014 Workshop on Analysis of Large Scale Social Interaction in MOOCs, pp. 60–65 (2014)
3. Yang, D., Sinha, T., Adamson, D., Rose, C.P.: Turn on, tune in, drop out: anticipating student dropouts in massive open online courses. In: Proceedings of NIPS Work Data Driven Education, pp. 1–8 (2013)
4. Jordan, K.: MOOC completion rate: the data (2013)
5. Ye, C., Biswas, G.: Early prediction of student dropout and performance in MOOCs using higher granularity temporal information. J. Learn. Anal. **1**, 169–172 (2014)
6. Coates, A., et al.: Text detection and character recognition in scene images with unsupervised feature learning. In: Proceedings of International Conference Document Anal. and Recognition ICDAR, pp. 440–445 (2011)
7. Wen, M., Yang, D., Ros, C.P., Rosé, C.P., Rose, C.P.: Linguistic reflections of student engagement in massive open online courses. In: Proceedings of 8th International Conference of Weblogs Social Media, ICWSM 2014, pp. 525–534 (2014)
8. Wen, M., Yang, D., Rosé, C.P.: Sentiment Analysis in MOOC Discussion Forums: What does it tell us? In: Proceedings of the 7th International Conference on Educational Data Mining (EDM), pp. 1–8 (2014)
9. Gardner, J., Brooks, C.: Student success prediction in MOOCs. User Model. User-Adapt. Inter. **28**, 127–203 (2018)
10. Hong, B., Wei, Z., Yang, Y.: Discovering learning behavior patterns to predict dropout in MOOC. In: 12th International Conference on Computer Science and Education, ICCSE 2017, pp. 700–704. IEEE. (2017)
11. Xing, W., Chen, X., Stein, J., Marcinkowski, M.: Temporal predication of dropouts in MOOCs: reaching the low hanging fruit through stacking generalization. Comput. Hum. Behav. **58**, 119–129 (2016)
12. Halawa, S., Greene, D., Mitchell, J.: Dropout prediction in MOOCs using learner activity features. In: Proceedings of the Second European MOOC Stakeholder Summit, pp. 58–65 (2014)
13. Sharkey, M., Sanders, R.: A process for predicting MOOC attrition. In: Proceedings of the EMNLP 2014 Workshop on Analysis of Large Scale Social Interaction in MOOCs, pp. 50–54 (2014)
14. Nagrecha, S., Dillon, J.Z., Chawla, N.V.: MOOC dropout prediction: lessons learned from making pipelines interpretable. In: International World Wide Web Conferences Steering Committee Proceedings of the 26th International Conference on World Wide Web Companion, pp. 351–359 (2017)
15. Bote-Lorenzo, M.L., Gómez-Sánchez, E.: Predicting the decrease of engagement indicators in a MOOC. In: Proceedings of the Seventh International Learning Analytics and Knowledge Conference on LAK 2017. pp. 143–147. ACM Press, New York (2017)
16. Liang, J., Yang, J., Wu, Y., Li, C., Zheng, L.: Big data application in education: Dropout prediction in Edx MOOCs. In: Proceedings of 2016 IEEE 2nd International Conference on Multimedia Big Data, BigMM 2016, pp. 440–443, IEEE (2016)
17. Chen, T., Guestrin, C.: Xgboost: A scalable tree boosting system. In: Proceedings of the 22nd ACM SIGKDD International Conference on Knowledge Discovery and Data Mining, pp. 785–794, ACM. (2016)
18. Dietterich, Thomas G.: Ensemble methods in machine learning. In: Kittler, J., Roli, F. (eds.) MCS 2000. LNCS, vol. 1857, pp. 1–15. Springer, Heidelberg (2000). https://doi.org/10.1007/3-540-45014-9_1

19. Ruipérez-Valiente, J.A., Cobos, R., Muñoz-Merino, P.J., Andujar, Á., Delgado Kloos, C.: Early prediction and variable importance of certificate accomplishment in a MOOC. In: Delgado Kloos, C., Jermann, P., Pérez-Sanagustín, M., Seaton, D.T., White, S. (eds.) EMOOCs 2017. LNCS, vol. 10254, pp. 263–272. Springer, Cham (2017). https://doi.org/10.1007/978-3-319-59044-8_31
20. Cristea, A.I., Alamri, A., Kayama, M., Stewart, C., Alshehri, M., Shi, L.: Earliest predictor of dropout in MOOCs: a longitudinal study of futurelearn courses. In: 27th International Conference on Information Systems Development (ISD) (2018)
21. Alshehri, M., et al.: On the need for fine-grained analysis of gender versus commenting behaviour in MOOCs. In: Proceedings of the 2018 The 3rd International Conference on Information and Education Innovations, pp. 73–77. ACM (2018)
22. Cristea, A.I., Alshehri, M., Alamri, A., Kayama, M., Stewart, C., Shi, L.: How is learning fluctuating? futurelearn MOOCs fine-grained temporal analysis and feedback to teachers and designers. In: 27th International Conference on Information Systems Development (ISD2018). Association for Information Systems, Lund (2018)
23. Dorfman, R.: A formula for the Gini coefficient. Rev. Econ. Stat. **61**, 146–149 (1979)
24. Friedman, J.H.: Greedy function approximation: a gradient boosting machine. Ann. Stat. **29**, 1189–1232 (2001)
25. Hastie, T., Rosset, S., Zhu, J., Zou, H.: Multi-class adaboost. Statistics and its. Interface **2**, 349–360 (2009)
26. Schapire, R.E., Freund, Y.: Boosting: Foundations and algorithms. MIT press, Cambridge (2012)
27. Breiman, L.: Random forests. Mach. Learn. **45**, 5–32 (2001)
28. An, S., Liu, W., Venkatesh, S.: Fast cross-validation algorithms for least squares support vector machine and kernel ridge regression. Pattern Recognit. **40**, 2154–2162 (2007)
29. Hinkley, D.V., Cox, D.: Theoretical Statistics. Chapman and Hall/CRC, London (1979)

Predicting Subjective Enjoyment of Aspects of a Videogame from Psychophysiological Measures of Arousal and Valence

Julien Mercier[1], Pierre Chalfoun[2,3(✉)], Matthew Martin[1,3], Ange Adrienne Tato[1,3], and Daniel Rivas[1,3]

[1] NeuroLab, University of Quebec in Montreal, Montreal, QC H2L 2C4, Canada
[2] User Research Lab, Ubisoft, Montreal, QC H2T 1S6, Canada
pierre.chalfoun@ubisoft.com
[3] La Forge, Ubisoft, Montreal, QC H2T 1S6, Canada

Abstract. The links between objective measures of affect and subjective ratings regarding a learning and performance episode are not well understood. Specifically, how a subjective appreciation is constructed from low-level affective reactions during a given experience remains largely unknown. The goal of this study is to investigate if the subjective appreciation of a videogame can be predicted from objective online measures of affect, namely arousal and valence. The participants were 35 undergraduate students with minimal experience with the first-person shooter genre of video games. They played FarCry Primal™, a first-person shooter-infiltration game, for 90 min. Results show that arousal, and not valence, is related to the subjective appreciation of a videogame. Since a continuous measure of arousal is cheap and relatively unobtrusive, the findings may have applications in the design of interactive computer applications, such as ITS and videogames, in helping pinpoint important segments in learning episodes that will affect a learner's subjective rating of her experience.

Keywords: Psychophysiology · Arousal · Valence · Subjective enjoyment · Videogames

1 Problem and Context

The goal of this study is to investigate if subjective appreciation of a videogame can be predicted from objective online measures of affect during key moments in gameplay. While subjective ratings are obtrusive and not always accurate, objective measures of affect are reductionist and may not capture the complexity of an affective experience [10]. In authentic contexts, in which an experience may extend over dozens of minutes, an affective judgement may be constructed from a few select events within the experience. Determining the basis for making a subjective judgment is key in improving such experiences, for example a learning or gaming experience.

A. Coy et al. (Eds.): ITS 2019, LNCS 11528, pp. 174–179, 2019.
https://doi.org/10.1007/978-3-030-22244-4_21

2 Theoretical Framework

Kreibig [7] has identified three classes of models of affective functioning. The first class is grounded in the autonomic nervous system. The second class focuses on brain-behavioral systems. The third class emphasizes psychological processes of meaning assessment (appraisal) and memory retrieval (as functions of associative networks). To meet our needs, work in a systems neuroscience perspective is needed to determine how to integrate these levels functionally. Circumplex models of affect [9], which derive specific emotions on the basis of two dimensions, arousal and valence, may be key in this endeavor. Since subjective judgments (interoception) of correlates of psychophysiological measures, such as valence and arousal, are often inaccurate [10], a complementary use of self-report measures should insist on the appraisal component of affect.

2.1 Electrodermal Activity and Arousal

From a psychophysiological perspective, it is suggested here that measuring aspects of affect can be done in an approach similar to cognitive load, as discussed for example by Antonenko et al. [2], in which temporally fine-grained psychophysiological measures is used to derive several indexes. Physiological correlates of emotions include cardiovascular, electrodermal and respiratory measures, supplemented with facial electromyography. According to [7], 36 indicators can be computed from these sources, including 20 for cardiovascular measures, 13 for respiratory measures and 3 for electrodermal measures. Concerning the three electrodermal measures, nonspecific skin conductance response rate (nSSR), and skin conductance level (SCL) and phasic response (SCR), we focus on the phasic response (SCR). SCR represents a variation in electrodermal activity at least partly related to a stimulus. SCR is commonly used as a measure of arousal as the properties of the signal and its underlying neurocognitive mechanisms are well-understood [4]. The required sensors are also relatively cheap and unobtrusive.

2.2 Frontal Asymmetry and Valence

Researchers have observed that the left and right hemispheres of the brain appear to be specialized for the expression and experience of positive and negative emotions, respectively [5]. Much of this evidence comes from EEG studies showing relative higher activity in left-frontal sites in response to positive stimuli, versus higher activity in right-frontal sites in response to negative stimuli. For example, Davidson et al. [3] found that left-frontal EEG leads had higher power when participants were watching a positive-emotion inducing television program, whereas right-frontal EEG leads had higher power when negative emotions were being elicited. Usually, the asymmetry in power is measured in the alpha band, and is referred to as frontal alpha asymmetry [11]. Because alpha power is inversely related to cortical activation, high relative alpha power in the left frontal region is reflective of less activation, and thus negative emotional states [8].

Earlier accounts suggested that frontal asymmetry is related to valence. However, more recent evidence suggests that frontal asymmetry is reflective of approach-avoidance motivation and emotions rather than valence, with greater relative left frontal activation indicating higher approach emotions, and higher right frontal activation indicating higher avoidance emotions [5, 11]. For example, studies manipulating the emotion of anger (a negatively valenced approach emotion) have found elicitations of anger from receiving insulting feedback produce greater left-frontal alpha activation compared to neutral feedback [6].

It is pertinent to highlight how alpha asymmetry is measured and how its measurement affects its interpretation. Typically, raw alpha powers of contralateral EEG sites (e.g., F3 and F4) are logarithmically transformed, and then a difference score is computed as a single measure of relative hemispheric activation. The logarithmic transformation reduces the skewness and kurtosis of the raw signal, and also mitigates individual differences in skull impedance between participants [1]. The index of alpha asymmetry is given by the equation (ln[right]-ln[left] alpha power), with a higher score reflecting relative greater left frontal activity (given that alpha is inversely related to cortical network activity). It is important to highlight that this is a relative measure of hemispheric activation and does not reflect the absolute activation in each individual hemisphere. For example, an increase in the ratio of activation can be driven entirely by a decrease of activation in the right hemisphere alone, and no change in the left hemisphere, somewhat contrary to the approach-avoidance theory [1]. While examining the individual contributions of both hemispheres individually is perhaps more informative in testing theories of cortical asymmetry, the ratio measure has consistently been found to be an effective measure of state-changes in approach-avoidance emotions and relatedly to high/low valence emotions [1, 11].

2.3 Research Questions

The theory reviewed predicts that a subjective experience that can be conscious and self-reported is constructed of objective affective events occurring at various moments in time during an experience. The research questions are as follows:

Is subjective enjoyment of specific weapons predicted from psychophysiological measures of arousal and valence during the use of these weapons in a shooter videogame?

Is subjective enjoyment of the shooter videogame in general predicted from psychophysiological measures of arousal and valence during the use of all weapons?

3 Method

3.1 Participants

The participants are 35 undergraduate students with homogeneous minimal experience with the first-person shooter genre of video games. They played FarCry Primal™ for 90 min, a first-person shooter game. After a thematic introduction video, the first missions are designed to scaffold the player in mastering the core mechanics of the game (moving, crafting, and camera control, interacting with enemies, managing the development of the character, managing resources, etc.). The ethics approval for this experiment was granted by the university of the first author.

3.2 Measures

Electrodermal activity and electroencephalography were recorded concurrently with the performance in the game (game events recorded in near real-time by the game console): 64-channel electroencephalography (EEG), and 1000 Hz electrodermal activity (EDA) are coupled with a screen capture from the game console.

Subjective enjoyment of the game and of its weapons was measured using four items presented as a five-point Likert scale (ranging from not fun at all to extremely fun). Three items concerned the specific weapons: "Overall, what did you think of the following weapons?" (bow, spear, club). One item concerned the game as a whole: "Overall, what did you think of Far Cry Primal?".

3.3 Data Preparation and Plan of Analysis

After signal decontamination, the EEG at sites F3 and F4 was subjected to the following transformation (ln[right]-ln[left] alpha power), providing two data points per second for valence. The EDA was downsampled to 10 Hz for ease of processing and then the SCR was taken as the measure of arousal. The measure of arousal was z-scored within-subject using a two-minute baseline collected before the gameplay while valence was z-scored within-subject over the entire episode. They were then segmented according to the gameplay, specifically regarding which weapon the character was using at a particular point in time. For the analyses, biometrics were averaged within players across relevant episodes. The research questions were answered by using multiple linear regression to predict enjoyment of game (enjoyment of: bow, club, spear, and whole game) from biometrics (valence and arousal).

4 Results

Table 1 presents the results for both research questions. They are addressed in turn next.

4.1 Is Subjective Enjoyment of Specific Weapons Predicted from Psychophysiological Measures of Arousal and Valence During the Use of These Weapons in a Videogame?

For all three weapons, arousal predicts appreciation whereas valence does not predict appreciation. Arousal is directly related to the appreciation of the club, indicating that relatively high arousal during use of the club increases its appreciation. In contrast, arousal is inversely related to the appreciation of the bow and spear, meaning that relatively low arousal during the use of these weapons increases their respective appreciation.

Table 1. Enjoyment of specific weapons and the whole game predicted by valence and arousal.

Model	β	*t*	Sig.
Bow			
Valence	.002	.077	.939
Arousal	−.109	−5.404	.000
Club			
Valence	−.003	−.142	.887
Arousal	.053	2.521	.012
Spear			
Valence	−.005	−.167	.867
Arousal	−.074	−2.428	.015
Game in general			
Valence	−.012	−.576	.564
Arousal	−.026	−1.308	.191

4.2 Is Subjective Enjoyment of the Videogame in General Predicted from Psychophysiological Measures of Arousal and Valence During the Use of All Weapons?

For predicting the appreciation of the game in general, valence and arousal during the identified weapon use episodes were far from reaching statistical significance.

5 Discussion

The goal of this study was to investigate if subjective appreciation of a videogame can be predicted from objective online measures of two main dimensions of affect: arousal and valence. The results do not completely correspond to the theory in the sense that only arousal, and not valence, was found to be predictive of subjective appreciation, contrary to current models stating that both dimensions are required to forge an emotion [10]. A possible explanation for our results is that a subjective appreciation is constructed from a subset of the objective experience, and that arousal may be the main driver in indexing specific episodes from the objective experience into a subjective experience.

The intriguing pattern of directionality between arousal and appreciation for specific weapons may be explained by strategic opportunities and choices within the game and may be elicited by the associated play style for this genre (melee vs ranged): with the club, the player may want to dispose of enemies quickly and risk heavy damage by facing enemies in close combat whereas the use of ranged weapons such as the bow and the spear affords for a more tactical and safe strategy.

Circumplex models of affect, in which arousal and valence are crossed to derive specific emotions, may help refine these results. This work will be extended by examining which emotions are best distinguished and which are the best combinations of objective and subjective indicators to characterize an emotional experience and to determine which specific emotions occurred during a specific time frame.

References

1. Allen, J.J.B., Coan, J.A., Nazarian, M.: Issues and assumptions on the road from raw signals to metrics of frontal EEG asymmetry in emotion. Biol. Psychol. **67**(1), 183–218 (2004). https://doi.org/10.1016/j.biopsycho.2004.03.007
2. Antonenko, P., Paas, F., Garbner, R., van Gog, T.: Using electroencephalography to measure cognitive load. Educ. Psychol. Rev. **22**, 425–438 (2010)
3. Davidson, R.J., Schwartz, G.E., Saron, C., Bennett, J., Goleman, D.J.: Frontal versus parietal EEG asymmetry during positive and negative affect. Psychophysiology **16**, 202–203 (1979)
4. Boucsein, W.: Electrodermal Activity, 2nd edn. Springer, New York (2012). https://doi.org/10.1007/978-1-4614-1126-0
5. Demaree, H.A., Everhart, D.E., Youngstrom, E.A., Harrison, D.W.: Brain lateralization of emotional processing: historical roots and a future incorporating "dominance". Behav. Cogn. Neurosci. **4**, 3–20 (2005)
6. Harmon-Jones, E., Sigelman, J.: State anger and prefrontal brain activity: Evidence that insult-related relative left-prefrontal activation is associated with experienced anger and aggression. J. Pers. Soc. Psychol. **80**(5), 797–803 (2001). https://doi.org/10.1037//0022-3514.80.5.797
7. Kreibig, S.D.: Autonomic nervous system activity in emotion: a review. Biol. Psychol. **84**, 394–421 (2010)
8. Reuderink, B., Mühl, C., Poel, M.: Valence, arousal and dominance in the EEG during game play. Int. J. Auton. Adapt. Commun. Syst. **6**(1), 45–62 (2013). https://doi.org/10.1504/IJAACS.2013.050691
9. Sander, D.: Models of emotion: the affective neuroscience approach. In: The Cambridge Handbook of Human Affective Neuroscience. Cambridge University Press, New York (2013)
10. Scherer, K.R.: What are emotions? And how can they be measured? Soc. Sci. Inf. **44**, 695–729 (2005)
11. Smith, E.E., Reznik, S.J., Stewart, J.L., Allen, J.J.B.: Assessing and conceptualizing frontal EEG asymmetry: an updated primer on recording, processing, analyzing, and interpreting frontal alpha asymmetry. Int. J. Psychophysiol. **111**, 98–114 (2017). https://doi.org/10.1016/j.ijpsycho.2016.11.005

Providing the Option to Skip Feedback – A Reproducibility Study

Amruth N. Kumar(✉)

Ramapo College of New Jersey, Mahwah, NJ 07430, USA
amruth@ramapo.edu

Abstract. Would providing choice lead to improved learning with a tutor? We had conducted and reported a controlled study earlier, wherein, introductory programing students were given the choice of skipping the line-by-line feedback provided after each incorrect answer in a tutor on if/if-else statements. Contrary to expectations, the study found that the choice to skip feedback did not lead to greater learning. We tried to reproduce these results using two tutors on if/if-else and switch statements, and with a larger subject pool. We found that whereas choice did not lead to greater learning on if/if-else tutor in this reproducibility study either, it resulted in decreased learning on switch tutor. We hypothesize that skipping feedback is indeed detrimental to learning. But, inter-relationships among the concepts covered by a tutor and the transfer of learning facilitated by these relationships compensate for the negative effect of skipping line-by-line feedback. We also found contradictory results between the two studies which highlight the need for reproducibility studies in empirical research.

Keywords: Skipping feedback · Worked example tutor · Reproducibility study

1 Introduction

Findings are mixed on the effect of choice on learning (e.g., [4, 5]). We had conducted a study on providing students of introductory programming courses the choice to skip feedback in a tutor on code-tracing [2]. We had noted two factors that could affect the outcome of the study:

- The programming concepts covered by the tutor were inter-related. The tutor provided line-by-line explanation of the correct solution in the style of worked examples [6] as feedback, and this explanation has been shown to improve learning [3]. Reading the feedback on one concept had the potential to help students also learn about other inter-related concepts. So, we expected students to be able to skip reading feedback on some problems without hampering their learning.
- The study was conducted while the tutor was being used for after-class assignment in introductory programming courses. In this unsupervised setting, some students may be motivated to maximize learning while others may be motivated to complete the assignment as quickly as possible. So, students may exercise the option to skip feedback for varying reasons - some related to learning, while others are not. Those who skip feedback for expediency may hamper their learning by skipping feedback.

A. Coy et al. (Eds.): ITS 2019, LNCS 11528, pp. 180–185, 2019.
https://doi.org/10.1007/978-3-030-22244-4_22

In the study, we found that providing choice did not lead to greater learning [2]. Students who had the choice to skip feedback needed marginally more problems to learn each concept, and their pre-post improvement was marginally less than that of those who did not have the choice. We tried to reproduce the results of this study.

Reproducibility is a core principle of scientific research. Reproducibility refers to the ability to draw the same results using different instruments, methods, protocols and/or participants [1]. The parameters of our reproducibility study were as follows:

- Instrument: Whereas the earlier study had used a single tutor on `if/if-else` statements, in this study, we used that tutor as well as a tutor on `switch` statements. Whereas `if/if-else` tutor presented only code-tracing problems, `switch` tutor presented problems on code-tracing as well as code-debugging. Both the tutors were adaptive, and presented feedback consisting of line-by-line explanation of the correct solution when a student's solution was incorrect.
- Subjects: In both the studies, the subjects were students in introductory programming courses. The earlier study was conducted in Fall 2015. The reproducibility study was conducted using `if/if-else` tutor in three subsequent semesters: Spring 2016–Spring 2017, and `switch` tutor in four semesters: Fall 2015–Spring 2017.

Both the studies were controlled, and used the same pretest-practice-post-test protocol. Experimental group subjects were given the choice to skip the line-by-line explanation feedback whereas control group students were not. Both the studies were conducted *in-natura*, i.e., under unsupervised conditions in real-life introductory programming courses where the tutors were used for after-class assignments.

2 The Reproducibility Study

Participants: The tutors on `if/if-else` and `switch` statements were used by students in introductory programming courses from multiple institutions that were randomly assigned to control or experimental group each semester. Table 1 lists the number of students in control group (no choice to skip feedback) and experimental group (choice to skip feedback) for the two tutors who granted IRB permission.

Table 1. Number of participants in the study under each treatment.

Tutor	Control group	Experimental group
`if/if-else`	528	322
`switch`	142	221

Instruments: The tutor on `if/if-else` statement presents code-tracing problems. In each problem, the student is asked to identify the output of a program containing one or more `if/if-else` statements, one output at a time, along with the line in the program that produced that output. If the student's answer is incorrect, the tutor provides line-by-line explanation of the correct answer [3]. The tutor covers 12 concepts on one-way (`if`) and two-way (`if-else`) selection statements.

The tutor on `switch` statement presents both code-tracing and code-debugging problems. In a code-debugging problem, a program containing a `switch` statement is presented and the student is asked to identify the line, code object and the specific syntax/semantic error applicable to the code object on the line. If the student's answer is incorrect, the tutor explains the genesis of the error contained in the program. The tutor covers 12 concepts.

Both the tutors cover C++, Java, and C#. Both are accessible over the web. They are part of a suite of problem-solving tutors for introductory programming topics called problets (www.problets.org). Typically, students use the tutors as after-class assignments, often multiple times till they have mastered all the concepts in the topic.

Protocol: Every time a software tutor is used, it administers pretest-practice-post-test protocol as follows:

- Pretest: During pretest, the tutor presents one problem per concept to prime the student model. If a student solves a problem correctly, no feedback is provided to the student. On the other hand, if the student solves a problem partially correctly, incorrectly, or opts to skip the problem without solving it, line-by-line explanation is presented to the student.
- Adaptive Practice: Once the student has solved all the pretest problems, practice problems are presented on only the concepts on which the student skipped solving the problem or solved the problem partially/incorrectly during pretest. On each such concept, the student is presented multiple problems until the student has mastered the concept, i.e., solved a minimum percentage (e.g., 60%) of the problems correctly. After each incorrectly solved problem, the tutor presents line-by-line explanation of the correct answer.
- Adaptive Post-test: During this stage, which is interleaved with practice, the student is presented a test problem on each concept already mastered by the student during practice.

Pretest, practice and post-test are administered back-to-back without interruptions, entirely over the web by the tutor. The entire protocol is limited to 30 min. Since this was a controlled study, experimental group had the option to skip the line-by-line explanation provided after the student had either skipped solving a problem or solved the problem incorrectly/partially, whereas control group did not. Students who skip solving the pretest problem on a concept or solve it partially/incorrectly, solve enough problems during practice to master the concept, and solve the post-test problem on the concept correctly are said to have **learned** the concept.

The grade on each code-tracing problem was normalized to $0 \rightarrow 1.0$. Code-debugging problems were graded as correct or incorrect (no partial grade). If a student used a tutor multiple times, we considered data from the session when the student had learned the most number of concepts. If the student did not learn any concepts, we considered data from the first session when the student had solved the most number of problems.

In order to account for the 30-min limit placed on each session, the variables of the study were designed to be insensitive to the number of problems solved. They were:

- Pretest score per problem to verify that the control and experimental groups were comparable;
- The time spent per pretest problem - to assess the impact of treatment on the pace of solving problems during pretest;
- The number of concepts learned as a measure of the amount of learning;
- The number of practice problems solved per learned concept, as a measure of the pace of learning. It was calculated by dividing the number of practice problems solved on all the learned concepts by the number of concepts learned;
- Pre-post change in grade per learned concept as a measure of improvement in learning.

The fixed factor was treatment: whether students did or did not have the option to skip line-by-line explanation.

3 Results and Discussion

`if/if-else` Tutor Results: One-way ANOVA analysis of the pretest score per problem and the time spent per pretest problem yielded no significant main effect for treatment. *So, the two groups were comparable.* Analysis of the number of concepts learned yielded no significant main effect for treatment. So, the *treatment did not lead to greater learning.* The number of practice problems solved per learned concept was not significantly different between the two groups. So, the treatment *did not affect the pace of learning.*

But, a significant difference was observed on the **pre-post change in score** on the learned concepts between control and experimental subjects [$F(1,348) = 5.797$, $p = 0.017$]: pre-post improvement was 0.844 ± 0.03 for control subjects as compared to 0.902 ± 0.038 for experimental subjects. *So, treatment led to greater improvement in score on learned concepts.*

`switch` Tutor Results: One-way ANOVA analysis of the pretest score per problem and the time spent per pretest problem yielded no significant main effect for treatment when only those who learned at least one concept were considered. *So, the two groups were comparable.*

Analysis of the **number of concepts learned** yielded significant main effect for treatment [$F(1,177) = 4.816$, $p = 0.03$]: among those who learned at least one concept, control subjects ($N = 65$) learned 2.877 ± 0.33 concepts whereas experimental subjects ($N = 133$) learned 2.416 ± 0.25 concepts. So, overall, the *option to skip feedback led to significantly less learning.*

The number of practice problems solved per learned concept was not significantly different between the two groups. So, the treatment *did not affect the pace of learning.* No significant difference was observed in the pre-post change in score on the learned concepts between control and experimental subjects.

Table 2 compares the results obtained in the earlier study with those from this reproducibility study. In the table, empty cells correspond to no significant difference found between treatments. Parenthesized results are only marginally significant.

Table 2. Comparison of the results from the two studies

Variable	`if/if-else` tutor		`switch` tutor
	Earlier study	Reprod. study	Reprod. study
Pretest time	Control > Exp.		
Concepts learned			Control > Exp.
Practice per concept	(Exp. > Control)		
Pre-post change	(Control > Exp.)	Exp. > Control	

The **time spent per pretest problem** was found to be significantly greater for control than experimental group in the earlier study. In the reproducibility study, even when the same `if/if-else` tutor was used, no significant difference was found between treatments. One explanation for this might be that experimental subjects in the reproducibility study skipped feedback on a small percentage of the solved problems: 9.09% (342 problems out of 3761 solved) in `if/if-else` tutor and 11.67% (501 problems out of 4295 solved) in `switch` tutor. So, even though they saved time when they skipped feedback, the time saved skipping feedback was small compared to the time the subjects spent solving problems.

No significant difference was found between treatments on the **number of concepts learned** in both the earlier study and the reproducibility study that used `if/if-else` tutor. But, in the reproducibility study that used `switch` tutor, experimental subjects learned significantly less than control subjects. One explanation might be that the concepts on `switch` statements are less inter-related than those on `if/if-else` statements: whereas `if/if-else` tutor presented only code-tracing problems, `switch` tutor presented both code-tracing and debugging problems, with no conceptual overlap between the two types of problems. So, there was less transfer of learning among concepts on `switch` than on `if/if-else`. Therefore, when students skipped feedback and did not benefit as much from transfer of learning, they ended up learning less. If this explanation is true, skipping feedback is indeed detrimental to learning. But, *inter-relationships among the concepts covered by a tutor and the transfer of learning facilitated by these relationships compensate for the negative effect of skipping feedback.* This is a hypothesis worth testing in the future.

In the earlier study, experimental subjects solved marginally more **practice problems per learned concept**. But, in this reproducibility study, we did not find any difference between treatments with either tutor. Since the result of the earlier study was only marginally significant, it may have been an artifact of the student population that should have been ignored in the previous study.

The final difference in Table 2 is on **pre-post change in score on learned concepts**. Whereas pre-post increase in score was marginally greater for control group subjects in the earlier study, it was significantly greater for experimental group subjects in the reproducibility study *that used the same* `if/if-else` tutor. The two results are clearly contradictory. One possible explanations for the contradictory results is that data selection criteria differed between the two studies. In the previous study, when a student used the tutor multiple times, data was considered from "only the first time when the student had solved all the pretest problems." In the current study, the session in which the student had learned the most concepts was chosen. If a student skips solving a pretest problem, it is marked as not attempted and would make the session less likely to be selected according to the criterion used in the previous study. The student could still go on to solve practice problems and learn several concepts, which would make the session more likely to be selected in the current study. Yet, we do not expect this difference in criteria to have affected more than a handful of students: usually, the more problems a student solved, the more concepts the student learned.

While the reason behind the contradictory results remains to be investigated further, the contradictory results themselves highlight the need for reproducibility studies in empirical research. Reproducibility studies might open up new research questions, expose nuances that the original work may not have considered, or buttress earlier results with additional empirical support.

Based on the earlier study and this reproducibility study, we conclude that providing the option to skip feedback does not increase learning. On the other hand, if the concepts covered by the tutor are not inter-related, the option to skip feedback will result in decreased learning.

Acknowledgments. Partial support for this work was provided by the National Science Foundation under grant DUE-1432190.

References

1. Drummond, C.: Replicability is not reproducibility: nor is it good science. In: Proceedings of Evaluation Methods for Machine Learning Workshop, 26th ICML, Montreal, Canada (2009)
2. Kumar, A.N.: Providing the option to skip feedback in a worked example tutor. In: Micarelli, A., Stamper, J., Panourgia, K. (eds.) ITS 2016. LNCS, vol. 9684, pp. 101–110. Springer, Cham (2016). https://doi.org/10.1007/978-3-319-39583-8_10
3. Kumar, A.N.: Explanation of step-by-step execution as feedback for problems on program analysis, and its generation in model-based problem-solving tutors. Technol. Instr. Cogn. Learn. (TICL) **4**(1), 65–107 (2006). Special Issue on Problem Solving Support in Intelligent Tutoring Systems
4. Meyer, B.J.F., et al.: Web-based tutoring of the structure strategy with or without elaborated feedback or choice for fifth- and seventh-grade readers. Read. Res. Q. **45**(1), 62–92 (2010)
5. Ostrow, K.S., Heffernan, N.T.: The role of student choice within adaptive tutoring. In: Conati, C., Heffernan, N., Mitrovic, A., Verdejo, M.F. (eds.) AIED 2015. LNCS (LNAI), vol. 9112, pp. 752–755. Springer, Cham (2015). https://doi.org/10.1007/978-3-319-19773-9_108
6. Sweller, J., Cooper, G.A.: The use of worked examples as a substitute for problem solving in learning algebra. Cogn. Instr. **2**, 59–89 (1985)

Reducing Annotation Effort in Automatic Essay Evaluation Using Locality Sensitive Hashing

Tsegaye Misikir Tashu[1,2](✉), Dávid Szabó[1], and Tomáš Horváth[1,3]

[1] Faculty of Informatics, Department of Data Science and Engineering, Telekom Innovation Laboratories, ELTE-Eötvös Loránd University, Pázmány Péter sétány 1117, Budapest, Hungary
{misikir, tomas.horvath}@inf.elte.hu, david@szdavid.hu
[2] Faculty of Informatics, 3in Research Group, ELTE-Eötvös Loránd University, Martonvásár, Hungary
[3] Faculty of Science, Institute of Computer Science, Pavol Jozef Šafárik University, Jesenná 5, 040 01 Košice, Slovakia

Abstract. Automated essay evaluation systems use machine learning models to predict the score for an essay. For such, a training essay set is required which is usually created by human requiring time-consuming effort. Popular choice for scoring is a nearest neighbor model which requires on-line computation of nearest neighbors to a given essay. This is, however, a time-consuming task. In this work, we propose to use locality sensitive hashing that helps to select a small subset of a large set of essays such that it will likely contain the nearest neighbors for a given essay. We provided experiments on real-world data sets provided by Kaggle. According to the experimental results, it is possible to achieve good performance on scoring by using the proposed approach. The proposed approach is efficient with regard to time complexity. Also, it works well in case of a small number of training essays labeled by human and gives comparable results to the case when a large essay sets are used.

Keywords: Locality Sensitive Hashing · Automatic essay scoring · Similarity search

1 Introduction

The introduction of Automatic Essay Evaluation (AEE) opened a new area for scoring and evaluation of essays using computers. Scoring an essay is time consuming and expensive for a human grader. For this reason, numbers of AEE systems have been developed. The research on AEE is ongoing for more than a decade, utilizing Machine Learning (ML) and Natural Language Processing (NLP) techniques. Most of the state-of-the-art AEE systems rely on supervised ML approaches which require huge amount of annotated training essays to train the scoring engine. There are cases where finding labeled training essay is difficult and requires high efforts to create. For high-stakes essay scoring, the effort that goes into manually labeling a large number of essays in

A. Coy et al. (Eds.): ITS 2019, LNCS 11528, pp. 186–192, 2019.
https://doi.org/10.1007/978-3-030-22244-4_23

order to train a good model might be justified, but it becomes difficult in settings where new essays are generated more frequently [1]. It is mainly because it is hard to decide that which and how many of the newly generated essays have to be labeled manually to train the scoring engine. Clustering techniques were used to choose essays to be labeled manually by the user [2, 3]. The rationale behind labeling training essays using clustering is that, after applying clustering, similar essays will end up in the same cluster. The annotator (assessor) will only score the centroid of each cluster and the score will be then propagated to all members of the cluster. The problem with such approach is that essays assigned to the same cluster are not absolutely similar. They are grouped into the same cluster based on their relative closeness to the centroid of the given cluster. Therefore, scoring only the centroid and propagating the score to all members of the cluster would make the scoring engine biased towards the centroid of the cluster. A more refined approach would be to manually label a small subset of representative essays (containing, among others, also the above-mentioned cluster prototypes) and utilize nearest neighbor method for predicting the score for a new essay. The problem of nearest neighbor method is that it requires huge computation effort to select the nearest neighbors to a given essay. In this work, we propose and implement an approach to enhance the efficiency of a nearest neighbor method using Locality Sensitive Hashing (LSH). Experimental evaluation on real-world data shows that the proposed approach is efficient and works well also in case of small sized training essay sets annotated by human. Little work has been done in investigating the extent to which the AEE performance depends on the availability of training data and what proportion of essays should be scored by the teacher manually, an issue what is concerned in this paper. The rest of this paper is organized as follows: Sect. 2 introduces the proposed model, Sect. 3 provides an overview of the existing works. Experiments and results are described in Sect. 4. Section 5 concludes the paper and discusses prospective plans for future work.

2 Local Sensitivity Hashing (LSH)

The most common and simplest way to find similar documents among the large number of documents is by mutually comparing all the documents. But comparing all pairs of documents is time-consuming and is not computationally efficient. To solve this problem, we should concentrate only on the pairs of documents which are likely to be similar rather than checking every pair. LSH is one of the useful methods used to address cases like this. LSH, belonging to a family of randomized algorithms, allows us to quickly find similar documents from large collection of documents. When finding and grouping similar essays, from a large database of essays, into some group, the processing time grows linearly with the number of documents and their complexity. In the process of finding similar items, LSH can substantially reduce the computational time at the cost of only a small probability of failing to find the absolute closest match [4].

A general idea behind LSH is to hash each document for several times in a way that similar documents with very high similarity are hashed into the same bucket. Then, we consider each pair related to the same bucket as a candidate pair in each hash. It is enough to only consider the candidate pairs to find the similar elements as depicted in the Fig. 1. The procedure for indexing essays in the essays set E using LSH is done as follows [5]: First, select k hash functions h randomly and uniformly from the LSH hash function family H and create L buckets for each hash function. Then, create a hash table by hashing all essays e in the essays set E into different buckets based on their hash values. When a new essay e' arrives, one is using the same set of k hash functions h to map e' into L buckets, one from each hash table. Then, retrieve all essays e from the L buckets and collect (join) them into a candidate set C. Lastly, for each essay e in C compute its distance to e' and assign the score of e to the score of that e' which has the smallest distance from e'. The probability that two essays $e1$ and $e2$ are hashed into the same bucket is proportional to their distance u, as defined in the Eq. 1.

$$P(u) = \Pr\left[h(e1) - h(e2\right] = \int_0^{w} \left(\frac{1}{u} fs\left(\frac{t}{u}\right)\left(1 - \frac{t}{w}\right)\right) dt \tag{1}$$

where fs is the probability density function of the hash H and w is the bucket width. For any given bucket width w, this probability decreases as the distance u increases.

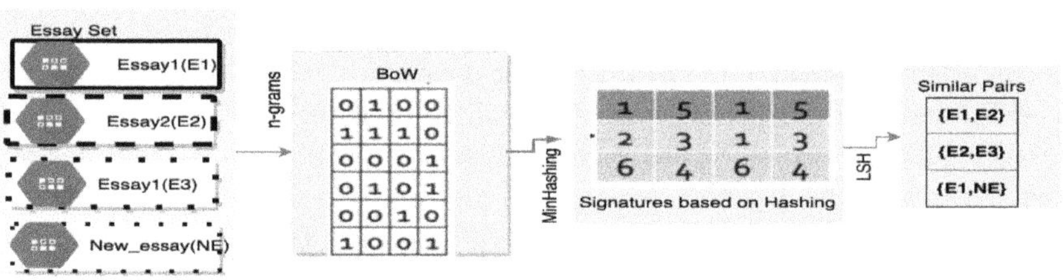

Fig. 1. An overview of the LSH method in searching candidate similar pairs of essays.

3 Related Work

The work done by Horbach et al. [6] investigates approaches for selecting the optimal number of short essays to be scored by a human grader and, later, to be used for training the scoring engine using clustering methods. After the human annotator labeled the optimal number of responses from each cluster, their approach propagates the human score to the rest of the cluster members. The distribution of scores in their dataset is unclear and they did not report agreement measures between the human score and the predicted score. Brooks et al. [2] proposed a more similar approach to the one of [6] that uses clustering to group student responses into clusters and sub-clusters. It allows the teachers access to these groupings, allowing them to read, grade, and provide feedback to large numbers of responses at once.

Basu et al. [7] also used clustering approaches that automatically find groupings and sub-groupings of similar answers for the same question using k-medoid and Latent Dirichlet Allocation (LDA). Zesch et al. [3] carried out an experiment of sample selection based on the output of clustering methods. By clustering the essays automatically, the items which are close to their centroids are labeled and added to the training data. By training on lexically diverse instances, the classifier should learn more than if trained on very similar instances. LSH has been used in k-nearest neighbor (KNN) based classification and clustering of large textual documents. In this work, we also investigate the possibility of selecting small proportion of training essays to be labeled which will result in comparable performance with engines trained on large number of essays.

4 Experimental Setup

In the experiments, we simulated a scenario when a new essay is being scored using the 1-nearest neighbor method assigning the score of the nearest essay to the given essay. In case there are more essays in the same distance to the new essay, their average is assigned as a score. We used the complete search for finding the nearest neighbors as a baseline in order to compare the efficiency of the proposed approach.

4.1 Dataset

The experiment was carried out on ten essay sets provided by the Hewlett Foundation at Kaggle[1] competition for AEE. All the datasets were rated by two human raters. Each essay is labeled with a numeric score from 0 to 2 and 3 and the answer length ranges from single phrases to several sentences. The following tasks were performed during preprocessing: tokenization; removing punctuation marks, determiners, and prepositions; transformation to lower-case; stopword removal and word lemmatization.

4.2 Evaluation Metrics and Parameter Settings

The machine score of each essay was compared with the human score to test the reliability of the proposed approach. Normalized root mean squared error (nRMSE) is used to evaluate the agreement between the predicted scores given by the proposed LSH-based algorithm and the actual human scores [8]. Rather than reporting the error, we will report the accuracy defined as.

$$Accuracy = 1 - nRMSE(ES) = 1 - \sqrt[2]{\frac{\sum_{e \in ES}(r(e) - h(e))^2}{ES}} \tag{2}$$

where *ES* is the essay set used, *e* denotes an essay, *r(e)* and *h(e)* are referring to the predicted rating and the human rating, respectively, for *e*. Rating here means how the essay is similar to the reference essay. The performance of the LSH algorithm depends on the similarity threshold, the number of permutations and the window size (the length of n-grams). The Jaccard similarity was used with a similarity threshold set to 0.8. The *MinHash* LSH is used and will be optimized for the given threshold by minimizing the false positive and false negative rates; The number of permutations used by the

[1] https://www.kaggle.com/c/asap-sas

MinHash to be indexed were set from the interval [10,128]; Window size of 2 and 3 were used; The proportion of train set to the whole set was set from the interval [20%, 40%].

4.3 Results and Discussion

After tuning the parameters of the proposed LSH approach, its performance w.r.t. the different values of its parameters are presented in Table 1. The results in Fig. 2 shows the percentage of unknown values at window size 2 (left) and at window size 3 (right), while the number of permutations is increasing from 10 to 128, and, the proportion of train dataset is also increasing from 20% to 40%. The number of essays for which LSH was unable to find any neighbor during similarity search are called unknowns and show the stochastic nature of LSH algorithm: with every run, there are unknowns based on the actual random parameters of the hash functions.

Table 1. Accuracy of the proposed LSH method with regard to different parameter settings.

No_perm	Accuracy at window size = 2			Accuracy at window size = 3		
	20%	30%	40%	20%	30%	40%
10	0.8942	0.8963	0.8940	0.8820	0.8840	0.8807
30	0.8980	0.8970	0.8991	0.87520	0.8727	0.8700
50	0.8973	0.8970	0.8977	0.8816	0.8780	0.8754
70	0.8991	0.8992	0.8972	0.8626	0.8625	0.8646
90	0.8986	0.8989	0.8998	0.8698	0.8713	0.8643
128	0.8990	0.8993	0.8975	0.8605	0.8578	0.8716

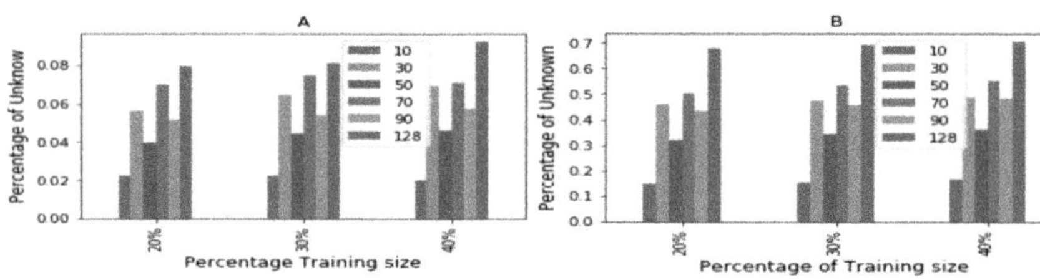

Fig. 2. The percentage of unknowns (left with window size = 2 and right window size = 3).

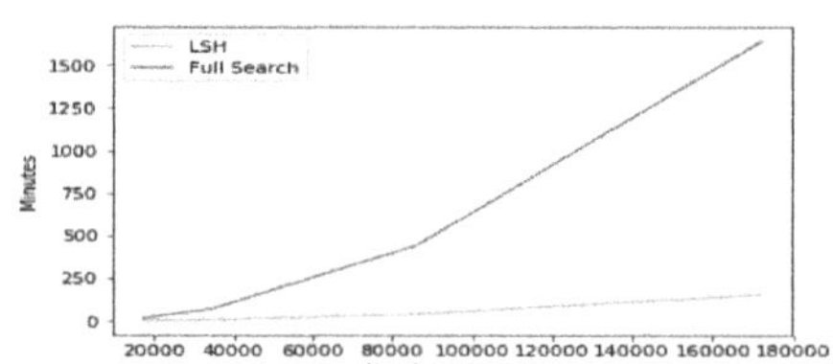

Fig. 3. Runtime using LSH and Full search.

When the window size is increasing from 2 to 3, the number of unknowns is increasing in very high rate. The lowest value of unknown with window size 2 is 0.020 and the highest is 0.092 while 0.15 is the lowest and 0.704 is the highest values of unknown at window size 3. After comparing the accuracy and the number of unknown at different values of the LSH parameters, the best accuracy which is 89.83% and the lowest percentage of unknown 2.2% was achieved where number of permutations is equal to 10 and similarity threshold equals to 80% by using only 30% of annotated training essay set. As the number of unknowns are increasing with the increase of window size, we did not report the accuracy of LSH for window size greater than 3. As we can see in the Fig. 3, LSH is also computationally more efficient than the full search method. From our experimental results, we can deduce that nearest neighbor-based AEE engines can work well with small training essay set and they can perform comparably to those engines using relatively high training essay set. This way we can decrease the annotation efforts required to create training essay set and also biases during annotation due to human annotation.

5 Conclusions

A reliable AEE engine requires large amount of labeled training data for the scoring engine to work very well. Labeling large amount of training essays is, however, time-consuming and exhausting for a human rater. In this work, we proposed and implemented a new way of selecting small size of training data set to be annotated that will be used to train the scoring engine using LSH. The performance of LSH-based AEE system was evaluated and compared at different values of its parameters with respect to accuracy and percentage of unknown essays. The proposed approach shows good performance and it is worth further investigation and development.

References

1. Heilman, M., Madnani, N.: The impact of training data on automated short answer scoring performance. In: Tenth Workshop on Innovative Use of NLP for Building Educational Applications, pp. 81–85 (2015)
2. Brooks, M., Basu, S., Jacobs, C., Vanderwende, L.: Divide and correct: using clusters to grade short answers at scale. In: The First ACM Conference on Learning @ Scale Conference, pp. 89–98. ACM, New York (2014)
3. Zesch, T., Heilman, M., Cahill, A.: Reducing annotation efforts in supervised short answer scoring. In: Tenth Workshop on Innovative Use of NLP for Building Educational Applications, pp. 124–132 (2015)
4. Slaney, M., Casey, M.: Locality-sensitive hashing for finding nearest neighbors [Lecture Notes]. IEEE Signal Process. Mag. **25**, 128–131 (2008)
5. Kim, Y.B., Reilly, U.O.: Large-scale physiological waveform retrieval via locality-sensitive hashing, pp. 5829–5833 (2015)

6. Horbach, A., Palmer, A., Wolska, M.: Finding a tradeoff between accuracy and rater's workload in grading clustered short answers. In: The 9th Language Resources and Evaluation Conference (LREC), pp. 588–595 (2014)
7. Basu, S., Jacobs, C., Vanderwende, L.: Powergrading: a clustering approach to amplify human effort for short answer grading. Trans. ACL (2013)
8. Misikir Tashu, T., Horvath, T.: Pair-wise: automatic essay evaluation using word mover's distance. In: 10th International Conference on Computer Supported Education, CSEDU, vol. 2, pp. 59–66. SciTePress (2018)

Representing and Evaluating Strategies for Solving Parsons Puzzles

Amruth N. Kumar(✉)

Ramapo College of New Jersey, Mahwah, NJ 07430, USA
amruth@ramapo.edu

Abstract. Parsons puzzles are popular for programming education. Identifying the strategies used by students solving Parsons puzzles is of interest because they can be used to determine to what extent students use the strategies typically associated with programming expertise, and to provide feedback and monitor the progress of students in a tutor. We propose solution sequence as an approximation of the student's strategy for solving Parsons puzzles. It is scalable in terms of both the size of the puzzle and the number of students solving the puzzle. We propose BNF grammar to represent desirable puzzle-solving strategies associated with expert programmers. This representation is extensible and agnostic to the puzzle-solving strategies themselves. Finally, we propose a best match parser that matches a student's solution sequence against the BNF grammar of a desirable strategy and quantifies the degree to which the student's solution conforms to the desirable strategy. As a proof of concept, we used the parser to analyze the data collected by a Parsons puzzle tutor on if-else statements over five semesters and found a significant difference between C++ and Java puzzle-solvers in terms of their conformance to one desirable puzzle-solving strategy. Being able to tease out the effects of individual components of a strategy is one benefit of our approach: we found that relaxing shell-first constraint in the strategy resulted in significant improvement in the conformance of both C++ and Java students.

Keywords: Parsons puzzle · Puzzle-solving strategy · Context free grammar · Evaluation

1 Introduction

Parsons puzzles were first proposed to "provide students with the opportunity for rote learning of syntactic constructs" in Turbo Pascal [13]. In a Parsons puzzle [13], the student is presented a problem statement, and the program written for it. The lines in the program are provided in scrambled order. The student is asked to re-assemble the lines in their correct order.

Parsons puzzles have been proposed for use in exams [2], since they are easier to grade than code-writing exercises, and yet, scores on Parsons puzzles correlate with scores on code-writing exercises [2]. Researchers have found solving Parsons puzzles to be part of a hierarchy of programming skills alongside code-tracing [11]. In electronic books, students have been found to prefer solving Parsons puzzles more than

A. Coy et al. (Eds.): ITS 2019, LNCS 11528, pp. 193–203, 2019.
https://doi.org/10.1007/978-3-030-22244-4_24

other low-cognitive-load activities such as multiple choice questions and high-cognitive-load activities such as writing code [4]. Solving Parsons puzzles was found to take significantly less time than fixing errors in code or writing equivalent code, but resulted in the same learning performance and retention [3]. So, Parsons puzzles have been gaining popularity for use in introductory programming courses.

Each Parsons puzzle has only one correct solution. So, the correct solution, i.e., the final re-assembled program will be the same for all the students. However, the temporal order in which students go about assembling the lines of code will vary among the students. This order indicates their solution strategy influenced by their understanding of the syntactic and semantic relationships among the lines of code, e.g., assembling a declaration statement before an assignment statement; or assembling the entire shell of a control statement before filling in its contents.

Recently, there has been interest among researchers in identifying the solution strategies used by students when solving Parsons puzzles. One study on Python Parsons puzzles [7] observed that some students re-assembled the lines using "linear" order, i.e., the random order in which the lines were provided. These researchers also observed backtracking and looping behavior, which were unproductive. Another preliminary study on Python Parsons puzzles observed that experts used top-down strategy to solve the puzzles [8], i.e., function header first, followed by control statements and eventually, individual statements. In another study of a Python Parsons puzzle tutor [5], researchers found that one common strategy was to focus on types of program statements. Novices often grouped lines based on indentation, whereas experts often built the solution top-down, demonstrating a better understanding of the program model. This study used think-aloud protocol and audio/video recordings to identify puzzle-solving strategies of novices and experts.

The benefits of identifying puzzle-solving strategies of students are manifold. Research shows that the strategies used by novices are different from those used by experts for programming tasks [15]. Experts use a strategy of reading a program in the order in which it is executed, and this leads to the development of a hierarchical mental model [12]. Expert programmers show more evidence of using a hierarchical mental model when understanding a well-written program than novices [6]. Moreover, a novice's success in learning to program is influenced by the student's mental model of programming [1, 14]. So, identifying a student's puzzle-solving strategy helps:

- determine whether the student is using the strategies typically used by experts;
- provide feedback to nudge the student towards adopting the successful strategies used by experts with the expectation that doing so will help the student develop the mental models associated with programming expertise, and thereby become a better programmer himself/herself, which is the goal of using Parsons puzzles; and
- determine whether the puzzle-solving strategy of students improves with practice when they use a tutor.

In order to identify the solution strategies used by students when solving Parsons puzzles, we propose (1) **solution sequence** as an approximate linear representation of their puzzle-solving behavior; (2) **BNF grammar** as a flexible representation of desirable solution strategies; (3) a **best-match parser** that matches a student's solution sequence against the BNF grammar for the puzzle to quantify the student's conformance to a desirable solution strategy. As a proof of concept, we apply our approach to analyze the solution strategies of students in data collected by a Parsons puzzle tutor on `if-else` statements.

2 The Solution

2.1 Action Sequence, Inner Product and Solution Sequence

Think-aloud protocol has been used to identify puzzle-solving strategies [5, 8]. But, this protocol is expensive and not scalable to larger numbers of students. The other approach used to identify strategies is the conversion of interaction traces (log data) into state diagrams [7, 8], where states represent snapshots of the solution in progress. But, this approach leads to combinatorially explosive number of states for any puzzle that contains more than a few lines of code, and is hence, not scalable to larger puzzles.

Instead, we propose to represent the student's puzzle-solving behavior as the temporal order in which the lines of code in the solution are assembled in their correct spatial location, e.g., if the puzzle contains 5 lines, and the student starts by placing line 3 in its correct location, followed by lines 1, 5, 2 and 4 in their correct locations, we represent the puzzle-solving behavior of the student as the **solution sequence** [3, 1, 5, 2, 4].

Solution sequence is itself derived from the **action sequence** of the student, which is the sequence of actions carried out by the student to solve a Parsons puzzle. We make a simplifying assumption to generate the solution sequence from action sequence. A student may move a line of code multiple times in the process of solving the puzzle. But, in order to generate the student's solution sequence, we consider only the last time the student moves the line before arriving at the correct solution for the puzzle, e.g., suppose the student moves the lines of a puzzle containing 5 lines in the following order: 3, 5, 4, 2, 1, 5, 2, 4, which is the student's action sequence. The corresponding solution sequence is calculated as the **inner product** of the action sequence and the time of submission of the correct solution (shown in Fig. 1). The resulting solution sequence is [3, 1, 5, 2, 4] corresponding to the order in which each of the 5 lines was last placed in its correct location in the solution. This inner product transformation eliminates from the solution sequence backtracking and looping behavior found in action sequence – so, it loses information about unproductive behaviors of the student [7]. But, it retains information about the sequence of decisions the student takes about the final placement of those lines just as the correct solution falls into place. In other words, it captures the thinking of the student in putting the solution together after any trial-and-error activities such as backtracking and looping. It is a reflection of the syntactic and semantic relationships the student sees among the lines of code in the puzzle, i.e., it is a reflection of the student's mental model of the program. Therefore, it is an approximation of the puzzle-solving strategy of the student.

For a puzzle containing n lines of code, the length of action sequence is greater than or equal to n. The length of solution sequence is n. If a student solves a puzzle with no redundant actions, the solution sequence of the student is the same as the action sequence.

So, instead of considering combinatorially explosive number of states, we consider a solution sequence of linear complexity, whose size is the number of lines in the code. This solution sequence can be automatically extracted from the log data collected by the tutor as compared to manually eliciting it using think-aloud protocol. So, our approach scales both with the number of lines in the puzzle and the number of students solving the puzzle.

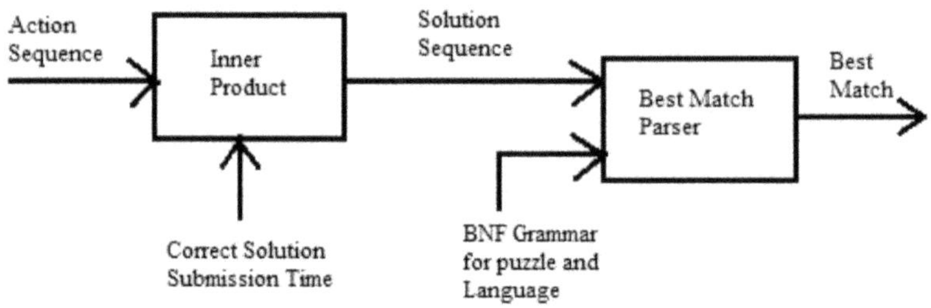

Fig. 1. Architecture of our approach

2.2 Desirable Solution Strategies

Experts have been found to use top-down strategy to solve Parsons puzzles [5, 8]. When reading a program, experts do so in the order in which it is executed, which leads to the development of a hierarchical mental model [12], a model that conceptualizes the elements of the program as forming a layered network of components, each of which can be of arbitrary depth and breadth [10]. Expert programmers also use a hierarchical mental model as compared to novices when understanding a program [6]. So, some desirable puzzle-solving strategies are those that use a top-down strategy and/or a hierarchical mental model of the program.

It is not clear that there is only one ideal strategy for solving Parsons puzzles – even experts have been seen to switch strategies [5]. Therefore, any approach for incorporating puzzle-solving strategies associated with programming expertise into Parsons puzzle tutors must accommodate a variety of strategies.

A generalized mathematical model of puzzle-solving strategies is characterized as follows: The lines in a puzzle may be grouped into subsets, such that all the lines within a subset must be assembled contiguously in time. The order among the subsets as well as among the lines within each subset may be total, partial or none. For example, two of the subsets within a puzzle may be: the set of declaration statements and the set of lines containing the shell of an `if-else` statement. A good strategy for solving the puzzle might require that declaration statements must be assembled before `if-else` statement (total order). The various lines of code within the set of declaration statements may be assembled in any order (no order). The braces within the shell of an `if-else` statement must be assembled in partial order: the closing brace must be assembled after the opening brace, but the braces of if-clause and else-clause may be assembled in either order.

Given these properties, we propose to use context free grammar or Backus-Naur Form (BNF) grammar to represent desirable puzzle-solving strategies. Once desirable strategies are represented using BNF grammar, a parser can be used to determine the degree to which the solution sequence of a student conforms to one or more desirable strategies. Since the BNF grammar can be used to simultaneously represent any number of strategies for a puzzle, whether they are complementary or contradictory, using it does not require commitment to any single idealized solution strategy.

2.3 BNF Grammar

A BNF grammar contains rules. Each rule is made up of terminals and non-terminals. Terminals are elements of the language, e.g., eat, drink. Non-terminals are types of the elements, e.g., noun, verb. Each rule contains a left hand side and a right hand side. The left hand side of a rule is a single non-terminal. The right hand side is a sequence of terminals and non-terminals. Alternatives in a rule are separated by vertical stroke.

In our case, the elements of the language are line numbers. The non-terminals are types of lines in the puzzle, e.g., declaration, output. Each rule specifies the temporal order in which a type of lines is assembled. Consider the following `if-else` statement to print two numbers in ascending order in C++:

```
1   if( first < second )
2   {
3      cout << first << endl;
4      cout << second << endl;
5   }
6   else
7   {
8      cout << second << endl;
9      cout << first << endl;
10  }
```

In the code, lines 1, 2, 5, 6, 7 and 10 constitute the shell of the `if-else` statement. Lines 3 and 4 are if-clause. Lines 8 and 9 are else-clause. A good programming practice is to assemble the entire shell of the `if-else` before assembling the if-clause and else-clause. This practice may be expressed in BNF grammar using the following rules:

<if-else> → <shell> <if-clause> <else-clause> | <shell> <else-clause> <if-clause>
<shell> → 1 2 5 6 7 10 | 1 6 2 5 7 10 | 1 6 7 10 2 5
<if-clause> → 3 4 | 4 3
<else-clause> → 8 9 | 9 8

In the grammar, non-terminals are enclosed in angle brackets <>. The first rule states that when assembling the `if-else` statement, the recommended practice is to assemble the entire shell first, followed by if-clause and else-clause in either order. Similarly, the third rule states that when assembling if-clause, lines 3 and 4 can be placed in their correct location in either order. According to the grammar, every possible combination of rules represents a desirable solution strategy. So, the above grammar represents 2 × 3 × 2 × 2 = 24 different solution strategies.

The grammar is extensible. If it is later determined that another good strategy would be to assemble the braces enclosing if-clause and else-clause at the same time as the clauses, adding the following rules to the grammar will accommodate the new strategy:

<if-else> → 1 <if-block> 6 <else-block> | 1 6 <if-block> <else-block>
<if-block> → 2 <if-clause> 5
<else-block> → 7 <else-clause> 10

Each BNF grammar is specific to a puzzle and programming language (See Fig. 1). So, if a tutor contains 10 puzzles and can be used for 3 different programming languages, 30 different BNF grammars must be encoded to analyze the data collected by the tutor.

2.4 A Best Match Parser

A parser takes a sentence and a grammar as inputs and outputs whether the sentence is correct according to the grammar. In our case, the sentence is a solution sequence. We not only wanted to know whether a solution sequence conformed to a grammar, but also the degree to which it conformed if it did not fully conform to the grammar. So, we developed a parser that takes two inputs: a BNF grammar and a solution sequence. It returns a number representing the **best match**, i.e., the maximum number of consecutive lines in the solution sequence that conform to any combination of rules in the grammar for the puzzle (See Fig. 1). It uses depth-first search to do so.

For example, given the grammar in Sect. 2.3, the behavior of the parser is as follows:

- The solution sequence [1, 2, 5, 6, 7, 10, 8, 9, 3, 4] is completely correct. So, the parser returns 10, the length of the solution sequence.
- The solution sequence [1, 6, 2, 5, 7, 10, 3, **9**, 4, 8] is correct up to line 9 - the student did not complete assembling if-clause before moving on to else-clause. The parser returns 7, the number of lines correct up to line 9.
- The solution sequence [1, 6, 7, 10, **5**, 2, 3, 4, 8, 9] is correct up to line 5 – the student assembled the closing brace before the opening brace enclosing if-clause in the `if-else` shell. The parser returns 4, the number of lines correct up to line 5.

Given a puzzle of n lines, the best match returned by the parser is in the range [0 … n].

3 A Proof-of-Concept Evaluation

As proof of concept, we evaluated the data collected by a Parsons puzzle tutor [9] (epplets.org) on `if-else` statements over 5 semesters: Fall 2016–Fall 2018. The tutor was used by students in introductory programming courses as after-class assignment.

Students used the tutor to solve puzzles in C++ or Java. Data for analysis was included from students who gave IRB permission for their data to be used in the study.

The first puzzle solved by all the students was on a program to read two numbers, and print the smaller value among them. Java version of the program is shown below. The program contains 14 manipulable lines in both C++ and Java: 2 lines of variable declaration (lines 9 and 11 below), 2 lines per input for 2 inputs (lines 13, 14 and 16, 17), followed by 8 lines of `if-else` statement (lines 19–26). The other lines in the program, such as comments, were provided *in-situ*.

```
1   // The Java program - 2005
2   import java.util.Scanner;
3   public class Problem
4   {
5      public static void main( String args[] )
6      {
7         Scanner stdin = new Scanner( System.in );
8         // Declare firstNum
9         int firstNum;
10        // Declare secondValue
11        int secondValue;
12        // Read firstNum
13        System.out.print( "Enter the first value");
14        firstNum = stdin.nextInt();
15        // Read secondValue
16        System.out.print( "Enter the second value");
17        secondValue = stdin.nextInt();
18        // Print the smaller value
19        if( firstNum  <  secondValue )
20        {
21           System.out.print( firstNum);
22        } // End of if-clause
23        else
24        {
25           System.out.print( secondValue);
26        } // End of else-clause
27     }  // End of method main
28  } // End of class Problem
```

The BNF grammar for the puzzle stipulated that it should be solved in the following order: the two declaration statements in any order, the two inputs in any order, `if-else` shell, followed finally by if-clause and else-clause in any order. The BNF grammar for Java is listed below.

<start-2005> → <declaration> <input> <output>
<declaration> → 9 11 | 11 9
<input> → <input1> <input2> | <input2> <input1>
<input1> → 13 14 | 14 13
<input2> → 16 17 | 17 16
<output> → <if-frame> <clauses>
<if-frame> → 19 <if-brace> 23 <else-brace> | 19 <if-brace> 23 |
19 23 <braces> | 19 23
<braces> → <if-brace> <else-brace> | <else-brace> <if-brace> |
<if-brace> | <else-brace>
<if-brace> → 20 22
<else-brace> → 24 26
<clauses> → <if-clause> <else-clause> | <else-clause> <if-clause>
<if-clause> → 21
<else-clause> → 25

We conducted one-way ANOVA with the best match as the dependent variable and programming language as the fixed factor. We found a significant main effect for programming language [$F(1,411) = 15.794$, $p < 0.001$]: C++ students had a best match score (5.62 ± 0.58, N = 114) significantly greater than Java students (4.24 ± 0.36, N = 298). The mean best match for C++ students corresponded to assembling declaration statements and both the inputs in the correct order. The same for Java students corresponded to assembling declaration statements and only the first input in the correct order. One possible explanation is that Java students were more likely to have been exposed to graphical user input, where inputs are separated spatially. So, they did not appreciate the linear order imposed on inputs in console input.

Our approach can be used to tease out the benefits of individual components of desirable strategies. We hypothesized that students might not appreciate the benefit of assembling the entire shell of an `if-else` statement before assembling the code contained in if-clause and else-clause. So, we added the following rules to the grammar that bypassed this restriction and allowed students to assemble the braces around if-clause and else-clause while assembling those clauses.

<output> → 19 <if-block> 23 <else-block> | 19 23 <if-block> <else-block>
<if-block> → 20 21 22 | 20 22 21 | 21 20 22
<else-block> → 24 25 26 | 24 26 25 | 25 24 26

We re-analyzed the data to calculate the best match with this extended grammar. ANOVA analysis with the original and extended best matches as the repeated measure and programming language as the fixed factor yielded a significant within-subjects effect for the change in grammar [$F(1,410) = 197.367$, $p < 0.0001$]: the mean best match increased from 4.932 ± 0.34 to 6.863 ± 0.56. As could be expected from earlier results, we found a significant between-subjects effect for language also [$F(1,410) = 18.599$, $p < 0.001$]. The best match of C++ students (N = 114) significantly improved from 5.623 ± 0.58 to 8.123 ± 0.955. Similarly, the best match of Java students (N = 298) significantly improved from 4.242 ± 0.36 to 5.604 ± 0.59. So, bypassing the requirement that the shell of `if-else` must be assembled completely before its contents resulted in a mean improvement in conformance of 2.5 lines for C++ students and 1.4 lines for Java students. *So, if assembling the shell first is indeed a good strategy, there is a need to better educate students about its importance.*

4 Discussion

We proposed **solution sequence** as an approximation of the student's strategy for solving Parsons puzzles. It is scalable in terms of both the size of the puzzle and the number of students solving the puzzle, compared to earlier approaches for modeling puzzle-solving strategies of students [5, 7, 8].

We proposed **BNF grammar** to represent desirable puzzle-solving strategies associated with expert programmers. This representation can accommodate complementary and contradictory strategies together, and is extensible. This representation scheme is agnostic to the puzzle-solving strategies themselves.

Finally, we proposed a best match parser that matches a student's solution sequence against the BNF grammar of a desirable strategy and quantifies the degree to which the student's solution conforms to the desirable strategy. As a proof of concept, we used the parser to analyze the data collected by a Parsons puzzle tutor on `if-else` statements over five semesters and found a significant difference between C++ and Java puzzle-solvers in terms of their conformance to one desirable puzzle-solving strategy. We demonstrated how our approach can be used to tease out the benefits of individual components of a desirable strategy. In the process, we also found that bypassing the constraint that shell must be assembled before its contents resulted in significant improvement in conformance of both C++ and Java students. Our approach can be used for any programming language or paradigm.

One shortcoming of our approach is that solution sequence is an approximation of the puzzle-solving strategy used by students because it loses information such as looping and backtracking behavior [7]. So, while it is suitable for understanding puzzle-solving strategies and mental models used by students, action sequence is better for diagnosing student misconceptions.

Our approach is designed to model desirable strategies and quantify how much students' solutions conform to those strategies, not to find patterns in students' solutions. Algorithms such as frequency counts and k-means clustering are better suited for finding patterns in students' solutions.

Currently, the best match parser returns the maximum number of consecutive lines in a solution sequence that match a desirable strategy. In the future, we will consider alternative matching algorithms such as Levenshtein algorithm.

In the future, we plan to use the BNF grammar representation of desirable puzzle-solving strategies, coupled with a generator (instead of a parser) to provide proactive hints to students as they solve puzzles in a tutor. Such hints might help students learn the puzzle-solving strategies associated with programming expertise and in turn, help students develop the mental models used by expert programmers and become better programmers themselves.

Acknowledgments. Partial support for this work was provided by the National Science Foundation under grants DUE-1432190 and DUE-1502564. The author thanks Anthony Bucci for suggesting the use of BNF grammars and Alessio Gaspar, Paul Wiegand and Jennifer Albert for associated discussions.

References

1. Cañas, J.J., Bajo, M.T., Gonzalvo, P.: Mental models and computer programming. Int. J. Hum Comput Stud. **40**(5), 795–811 (1994)
2. Denny, P., Luxton-Reilly, A., Simon, B.: Evaluating a new exam question: Parsons problems. In: Proceedings of the Fourth International Workshop on Computing Education Research (ICER 2008), New York, NY, USA, pp. 113–124. ACM (2008)
3. Ericson, B.J., Margulieux, L.E., Rick, J.: Solving Parsons problems versus fixing and writing code. In: Proceedings of the 17th Koli Calling International Conference on Computing Education Research (Koli Calling 2017), New York, NY, USA, pp. 20–29. ACM (2017)
4. Ericson, B.J., Guzdial, M.J., Morrison, B.B.: Analysis of interactive features designed to enhance learning in an Ebook. In: Proceedings of the 11th annual International Conference on International Computing Education Research (ICER 2015), New York, NY, USA, pp. 169–178. ACM (2015)
5. Fabic, G., Mitrovic, A., Neshatian, K.: Towards a mobile Python tutor: understanding differences in strategies used by novices and experts. In: Micarelli, A. (ed.) ITS 2016. LNCS, vol. 9684, pp. 447–448. Springer, Heidelberg (2016)
6. Fix, V., Wiedenbeck, S., Scholtz, J.: Mental representations of programs by novices and experts. In: Proceedings of the INTERACT 1993 and CHI 1993 Conference on Human Factors in Computing Systems (CHI 1993), New York, NY, USA, pp. 74–79. ACM (1993)
7. Helminen, J., Ihantola, P., Karavirta, V., Malmi, L.: How do students solve parsons programming problems? An analysis of interaction traces. In: Proceedings of the Ninth Annual International Conference on International Computing Education Research (ICER 2012), New York, NY, USA, pp. 119–126. ACM (2012)
8. Ihantola, P., Karavirta, V.: Two-dimensional Parson's puzzles: the concept, tools, and first observations. J. Inf. Technol. Educ.: Innov. Pract. **10**, 1–14 (2011)
9. Kumar, A.N.: Epplets: a tool for solving parsons puzzles. In: Proceedings of the 49th ACM Technical Symposium on Computer Science Education (SIGCSE 2018), New York, NY, USA, pp. 527–532. ACM (2018)
10. Letovsky, S.: Cognitive processes in program comprehension. In: Soloway, E., Iyengar, S. (eds.) Empirical Studies of Programmers, pp. 58–79. Ablex, Norwood (1986)
11. Lopez, N., Whalley, J., Robbins, P., Lister, R.: Relationships between reading, tracing and writing skills in introductory programming. In: Proceedings of the 4th International Workshop on Computing Education Research (ICER 2008), New York, NY, USA, pp. 101–112. ACM (2008)
12. Nanja, M., Cook, C.R.: An analysis of the online debugging process. In: Olson, G.M., Sheppard, S., Soloway, E. (eds.) Empirical Studies of Programmers: Second Workshop, pp. 172–184. Ablex, Norwood (1987)

13. Parsons, D., Haden, P.: Parson's programming puzzles: a fun and effective learning tool for first programming courses. In: Proceedings of the 8th Australasian Conference on Computing Education (ACE 2006), vol. 52, pp 157–163. Australian Computer Society, Inc. (2006)
14. Soloway, E., Ehrlich, K.: Empirical studies of programmer knowledge. IEEE Trans. Softw. Eng. **SE-10**(5), 595–609 (1984)
15. Winslow, L.E.: Programming pedagogy—a psychological overview. ACM SIGCSE Bull. **28**(3), 17–22 (1996)

Testing the Robustness of Inquiry Practices Once Scaffolding Is Removed

Haiying Li(✉), Janice Gobert, and Rachel Dickler

Rutgers University, New Brunswick, NJ 08901, USA
{Haiying.Li,Janice.Gobert,
Rachel.Dickler}@gse.rutgers.edu

Abstract. Intelligent tutoring systems (ITS) with simulated and virtual labs have been designed to enhance students' science knowledge, including content and inquiry practices; some systems do this via real-time scaffolding. Prior studies have demonstrated that scaffolding can benefit students' learning and performance. The present study aims to examine the robustness of scaffolding, delivered by a pedagogical agent by providing scaffolding on one activity, removing it, and then evaluating students' inquiry performance both over multiple time periods (in 40 days, 80 days, and 170 days) and across different topics, thereby addressing far transfer. 107 middle school students in grade 6 received adaptive scaffolding on the first inquiry topic (i.e. Animal Cell) in the intelligent tutoring system, Inq-ITS. Then they received no scaffolding for three topics, namely, Plant Cell, Genetics, and Natural Selection. Results showed that after removing scaffolding, students demonstrated continued growth of inquiry performance from time 1 to time 2, to time 3, and to time 4 for the practices of hypothesizing and collecting data, as well as from time 1 to time 2 and to time 4 for the practice of warranting claims. This pattern was not found in students' performance on the practice of interpreting data. These findings have implications for designers and researchers regarding the design of scaffolds for the NGSS' inquiry practices so that they can be effectively transferred. These data also point to the need for additional work to address content practice interactions.

Keywords: Science inquiry · Growth in inquiry performance · Scaffolding

1 Introduction

With the adoption of the Next Generation Science Standards [1] in many states, it is expected that students master disciplinary core ideas, crosscutting concepts, and scientific practices. Further, it is noted that engaging in scientific inquiry investigations is an appropriate and effective way to do this. However, without guidance and support, conducting scientific inquiry can be extremely challenging for students [2–4]. Specifically, students need scaffolds in order to meaningfully conduct science inquiry [5]. In brief, scaffolds are hints or structural guides provided to students in various forms in order to support them in accomplishing a task that may otherwise be beyond their present competency level [6]. Without scaffolding, it is difficult to be sure that students are following the necessary processes and gaining relevant

A. Coy et al. (Eds.): ITS 2019, LNCS 11528, pp. 204–213, 2019.
https://doi.org/10.1007/978-3-030-22244-4_25

understandings [7], for example, students may "fool around" [8], which can lead to alternative conceptions [9, 10].

1.1 Scaffolding in Science Inquiry Contexts

In the context of science inquiry, there are several different types of scaffolds, and these can be presented in multiple formats [11]. They may be presented in the form of pencil-paper materials [12, 13] or within virtual environments [14–17]. Paper-pencil scaffolds typically provide structural guidance to students to support particular steps of practices [18], such as hypothesizing and collecting data [12], and constructing explanations [13]. Scaffolds for these inquiry practices have also been developed for virtual environments, e.g., hypothesizing [14], conducting experiments [14, 15], analyzing and interpreting data [14, 15, 17], and constructing explanations [16]. Scaffolding inquiry has had varying degrees of success on students' learning and transfer across settings [19] or to new settings [11].

Some learning environments support students through *fixed scaffolding*, which is when the same amount of scaffolding is provided to students regardless of individual experience or performance [13, 16]. For instance, Tabak and Reiser's [16] inquiry software provided fixed scaffolding to students and was found to benefit students' content and process understandings. *Faded scaffolding* [12, 13, 17] can also be useful to learning; for example, in Co-Lab, an online system that gradually reduces the presence of explicit goals to guide students' inquiry, students benefited from the scaffolds, compared to controls, for the inquiry task of planning their inquiry investigations [17]. *Adaptive faded scaffolding*, in which the amount of scaffolding provided is individualized based on student's performance [11, 14], has been acknowledged as showing great promise in terms of supporting both learning and transfer [11]. Specifically, real-time adaptive scaffolds, the focus of our study here, can provide support based on students' specific needs, when they need it and when it is most effective for learning [20]. Additionally, this may also support transfer of inquiry practices to new settings [11], which we address here.

Real-time adaptive scaffolds have been implemented into Inq-ITS [21]. Inq-ITS is an intelligent tutoring system for middle school science in which students carry out inquiry investigations with microworld simulations in the domains of life, earth, and physical science. The real-time adaptive scaffolds in Inq-ITS support students on practices such as question formation/hypothesizing [22], carrying out investigations/data collection [23, 24], data analysis/interpretation, and warranting claims [25, 26] and all of the previously mentioned practices that are delivered by a pedagogical agent, Rex (scaffolds are currently being developed for the practice of constructing explanations in the claim, evidence, and reasoning format [27]).

In order to evaluate the effectiveness of these scaffolds, several studies [22–26] were conducted in which students were randomly assigned to receive scaffolding for each practice (i.e., data collection, data analysis, warranting claims). The data from these studies, analyzed using Bayesian Knowledge Tracing [28] and ANCOVAs, showed that students who received scaffolding were better able to both learn practices and transfer these competencies to new topics than were students who did not receive scaffolding [14, 22–26]). A recent study [29] demonstrated that students who receive scaffolding across practices improve on these practices at a quicker rate relative to students who do not receive scaffolding. Studies on the real-time, adaptive scaffolding in Inq-ITS, however, have yet to examine the robustness of adaptive scaffolding on inquiry competencies over time and across topics, i.e., once scaffolding is removed. This is the topic of the present study. We investigate whether adaptive scaffolding of inquiry practices on topic 1 is robust across topics at varying time intervals on future topics.

2 Method

2.1 Participants, Materials, and Procedure

107 6th grade students from a middle school in the northeastern United States participated in the present study. Of the population of students at the middle school, 39.2% are white, 20.6% are Hispanic, 23.5% are Asian, 11% are black, and 5.7% are two or more races.

In the present study, students completed virtual labs in Inq-ITS [22] in the following order: Animal Cell (3 microworld activities where students investigated how changing the number of organelles in the animal cell affected the health of the cell), Plant Cell (3 microworld activities where students investigated how changing the number of organelles in the plant cell affected the health of the cell), Genetics (3 microworld activities where students investigated how changing a mother monster's alleles impacted the traits of the monster's babies), and Natural Selection (4 microworld activities where students investigated how changing environmental factors impacted the presence of monsters with different traits).

Each Inq-ITS microworld activity contained four stages where students engaged in practices aligned to the NGSS [1]: forming questions/hypothesizing, carrying out investigations/collecting data, analyzing and interpreting data, and communicating findings. Adaptive, real-time scaffolding was available within the first three stages of the microworlds [22–26, 29] (scaffolding is currently being developed for the last stage of communicating findings in the claim, evidence, reasoning format [27]). These adaptive scaffolds were provided based on the automated scoring (described in more detail below) of students' performance on fine-grained components of inquiry practices.

If a student was identified as demonstrating low performance on a practice, the pedagogical agent (Rex) would pop-up on the student's screen with a speech bubble that oriented the student toward the particular inquiry practice he/she was completing (i.e., if a student is not running controlled trials when collecting data, then Rex might say, "*I think the data you're collecting won't help you test your hypothesis*"). If the student still demonstrates poor performance, Rex provides a procedural scaffold with hints on the steps the student should take to successfully engage in the practice (i.e., "*Design a controlled experiment by changing only the variable you are testing while keeping all the other variables the same*"). The student has the opportunity to request more information from Rex, in which case Rex provides a conceptual scaffold explaining components of the particular inquiry (i.e., "*Changing only the [IV] while keeping everything else the same lets you tell for sure if the [IV] affects the [DV]*"). Finally, if the student continues to struggle, Rex provides an instrumental scaffold with more direct instructions on how to successfully engage in the particular inquiry practice (i.e., "*Run pairs of trials where you: (1) Change only the [IV], and (2) Keep all the other variables the same*"). The student may not progress in the activity until he/she has addressed Rex's feedback. It is possible that Rex might not pop-up for a student at all during a scaffolded activity if the student demonstrates perfect performance on all inquiry practices on his/her first attempt. In the present study, students only had real-time, adaptive Rex scaffolding available in the first microworld topic (out of four) that they completed (i.e. Animal Cell).

2.2 Measures

The dependent variables in the present study were the students' scores on the four inquiry practices automatically assessed by knowledge engineered rules and educational data mining algorithms in the first three stages of the Inq-ITS system [22]. The four practices and their corresponding sub-components by which they were measured include (examples of correct sub-components are given for the Animal Cell: Vacuoles and Cell Storage virtual lab): (1) generating hypotheses, which was measured by identifying an IV (independent variable) and DV (dependent variable; i.e., vacuole size and cell storage), (2) collecting data, which was measured by testing the hypothesis, running pairwise targeted and controlled trials, and conducting a controlled experiment (i.e., running multiple trials where only the size of the vacuole was changed), (3) interpreting data, which was measured by correctly selecting the IV and DV for a claim, correctly interpreting the relationship between the IV and DV, and correctly interpreting the hypothesis/claim relationship (i.e., identifying that increasing the size of the vacuole increased the cell storage and if this conclusion matched the hypothesis), and (4) warranting claims, which was measured by warranting the claim with more than one trial, warranting with controlled trials, correctly warranting the relationship between the IV and DV, and correctly warranting the hypothesis/claim relationship (i.e., selecting at least two controlled trials with data showing that increasing the size of the vacuole increased the cell storage). Each inquiry practice subcomponent was automatically coded as 0 points if incorrect or 1 point if correct using the knowledge engineering and educational data mining techniques in Inq-ITS that have been validated in prior studies [22]. For the first activity (where students had the opportunity to receive scaffolding from Rex and reattempt inquiry practices), the analyses used students' performance on their first attempts for each inquiry practice before they received any scaffolding from Rex. The average score on each of the four inquiry practices across all of the activities within a topic (i.e. the average hypothesizing score across all 3 animal cell activities) was used for analyses.

One of the independent variables used in the present study was inquiry practice, which had four levels: hypothesizing, collecting data, interpreting data, and warranting claims. This study also had a variable for time of completion with four levels: at

Time 1 (December 2017), students completed the first Inq-ITS microworld topic (i.e. Animal Cell with Rex's support); at Time 2 (January 2018), students completed the second topic (i.e. Plant Cell); at Time 3 (March 2018), students completed the third topic (i.e. Genetics); and at Time 4 (June 2018) students completed the fourth topic (i.e. Natural Selection). There was a 40-day gap between completion of the first to second and second to third topics, and about 90 days between the completion of the third to last topic. The four inquiry practices and time of topic completion (i.e. first, second, third, and fourth time) were the two within-subject factors. The purpose of the study was to investigate whether there was any significant growth in and transfer of inquiry performance over time after removing the adaptive scaffolding provided only in the first topic.

3 Analyses, Findings, and Discussion

Repeated measures analyses were performed to investigate whether students' performance on each of the inquiry practices was robust after adaptive scaffolding was removed following completion of the first topic. Table 1 presents an overview of the means, standard deviations, minimum, and maximum scores.

Table 1. Statistics for inquiry practice × time of completion across four virtual labs (N = 107).

Time	Hypothesis		Data collection		Interpretation		Warranting	
	M(SD)	Min(Max)	M(SD)	Min(Max)	M(SD)	Min(Max)	M(SD)	Min(Max)
1 (*Day* = 1)	.79(.21)	.33(1.00)	.73(.26)	.00(1.00)	.82(.18)	.42(1.00)	.66(.27)	.00(1.00)
2 (*Day* = 40)	.91(.15)	.33(1.00)	.90(.19)	.00(1.00)	.88(.21)	.00(1.00)	.82(.28)	.00(1.00)
3 (*Day* = 80)	.86(.24)	.00(1.00)	.86(.24)	.00(1.00)	.77(.25)	.08(1.00)	.72(.28)	.00(1.00)
4 (*Day* = 170)	.90(.20)	.13(1.00)	.92(.14)	.38(1.00)	.81(.21)	.19(1.00)	.75(.24)	.25(1.00)

3.1 Performance on Inquiry Practices: Main Effect of Practices

Results of the repeated measures multivariate test for overall inquiry score was significant: $F(3, 104) = 26.59$, $p < .001$, $\eta^2 = .434$. Results for tests of within-subjects effects for practice was also significant: $F(3, 318) = 43.85$, $p < .001$, $\eta^2 = .293$. The pairwise comparisons of overall inquiry score (see Fig. 1) showed that students achieved higher scores on hypothesizing practice than interpreting data practice ($p < .001$, Cohen's $d = 0.27$) and warranting claims practice ($p < .001$, $d = 0.70$). Students' collecting data scores were also significantly higher than interpreting data ($p = .028$, $d = 0.18$) and warranting claims ($p < .001$, $d = 0.61$). Moreover, the interpreting scores were significantly higher than warranting claims scores ($p < .001$, $d = 0.44$). These results revealed that students' inquiry proficiencies vary with different practices. They have highest proficiency in hypothesizing and collecting data, followed by interpreting and warranting practices.

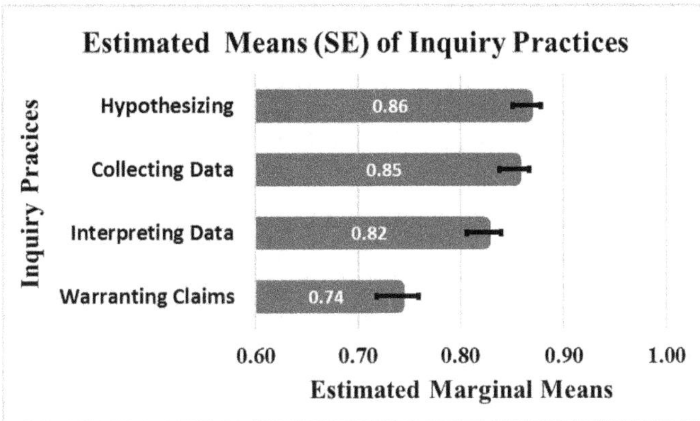

Fig. 1. Estimated marginal means and standard errors of four inquiry practice.

3.2 Growth over Time: Main Effect of Time

Results of the repeated measures analysis showed a significant multivariate effect for time, F (3, 104) = 20.93, $p < .001$, $\eta^2 = .376$. Results of the tests for within-subjects effects were also significant for time, F (3, 318) = 19.97, $p < .001$, $\eta^2 = .159$. The pairwise comparisons of overall inquiry performance over time showed that students achieved higher inquiry scores at Time 2 ($p < .001$, Cohen's $d = 0.68$), Time 3 ($p = .039$, Cohen's $d = 0.27$), and Time 4 ($p < .001$, Cohen's $d = 0.53$) relative to Time 1 (See Table 1). Results showed that inquiry scores were significantly lower at Time 3 than Time 2 ($p < .001$, Cohen's $d = 0.36$) (see left side of Fig. 2).

The topics completed at Time 1 and Time 2 were Animal Cells and Plant Cells, respectively, and therefore were similar in terms of the difficulty of content. Both topics are taught as part of the NGSS [1] middle school Life Science Strand 1 (From Molecules to Organisms: Structures and Processes). The increase in inquiry practice scores from Time 1 to Time 2 indicated that students benefitted from the adaptive scaffolding at Time 1 and demonstrated growth in inquiry performance, i.e., had further honed this practice at Time 2. The topics at Time 3 and Time 4 were more complex than the topics completed at Time 1 and Time 2, students still demonstrated growth in inquiry competencies at Time 3 and Time 4 relative to Time 1. However, at Time 3, students' inquiry performance decreased relative to Time 2. This drop in inquiry performance is likely explained by a change in the difficulty of topic (i.e. Plant Cell to Genetics) since genetics is one of the more difficult life science topics for middle and school students, requiring mathematical understandings of probability in addition to scientific content [30, 31]. Even though there is a decrease in scores from Time 2 to Time 3 due to the increase in the complexity of the topic (i.e. genetics), the performance at Time 3 still significantly improved relative to Time 1. Overall, these findings indicate robust effects of our scaffolding on students' inquiry practice competencies since adaptive scaffolding was removed after the first topic. The effect of scaffolding from Time 1 was successfully maintained over 40 days, 80 days, and 170 days for the overall inquiry practice score.

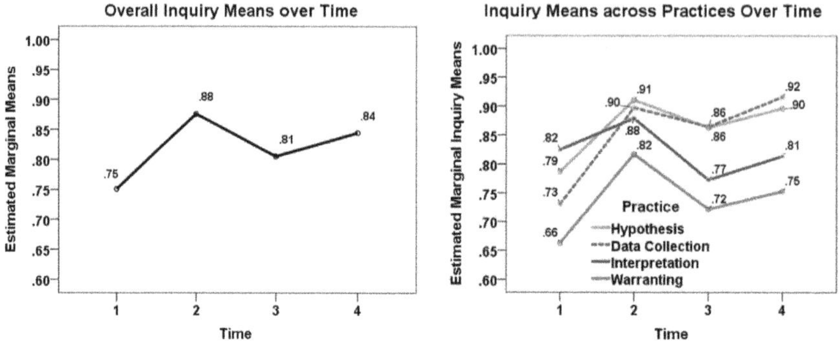

Fig. 2. Graph of overall average inquiry scores over time (Left) and mean score on each inquiry practice over time (Right).

3.3 Growth over Time: Interaction Between Time and Inquiry Practices

Results of the repeated measures multivariate analyses also showed a significant two-way interaction between time and inquiry practice: F (9, 98) = 11.00, $p < .001$, $\eta^2 = .503$. Results for tests of within-subjects effects were also significant for this interaction: F (9, 954) = 9.28, $p < .001$, $\eta^2 = .080$. Pairwise comparisons (see right side of Fig. 2) showed that for the practice of hypothesizing, students performed significantly higher at Time 2, $p < .001$, $d = 0.69$; Time 3, $p = .020$, $d = 0.34$; and Time 4, $p < .001$, $d = 0.54$, relative to Time 1. A similar pattern was found for the practice of collecting data, i.e., students performed significantly higher at Time 2, $p < .001$, $d = 0.73$, Time 3, $p < .001$, $d = 0.54$, and Time 4, $p < .001$, $d = 0.88$, relative to Time 1. This pattern was not found for the practice of interpreting data or for the practices of warranting claims. A different pattern emerged for the practice of warranting claims. Specifically, students achieved higher scores at Time 2, $p < .001$, $d = 0.56$, and Time 4, $p = .004$, $d = 0.35$, relative to Time 1 for the practice of warranting claims.

The findings for the specific practices of hypothesizing and data collection are similar to those found for the main effect of time. There was continuous growth in performance, i.e., a honing of practices, for hypothesizing and collecting data from Time 1 to Time 2, to Time 3, and to Time 4. These patterns demonstrate a growth in student performance for the practices of hypothesizing and collecting data regardless of the difficulty of topic from Time 1 to Time 4. For the practice of warranting claims, however, a significant increase was found only when comparing Time 1 to Time 2 and when comparing Time 1 to Time 4. For the practice of warranting claims, the difficulty of content (i.e. genetics) perhaps influenced performance at Time 3, where there was not a significant improvement in scores.

We found that students achieved better interpreting data performance at Time 2 than at both Time 3 ($p < .001$, $d = 0.46$) and Time 4 ($p = .009$, $d = 0.32$). We also found that students achieved higher warranting claims scores at Time 2 than at Time 3, $p = .003$, $d = 0.34$. The decreasing performance on interpreting data practice from Time 2 to Time 3 and from Time 2 to Time 4 is likely due to the increasing complexity of topics from plant cell (Time 2) to genetics (Time 3) and natural selection (Time 4). This phenomenon also occurred in warranting claims performance: decreasing from Time 2 to Time 3, likely due to the increase in difficulty of the topic (genetics). Another possibility is that these practices, namely, interpreting data and warranting claims, interact with content knowledge to a greater degree (than hypothesizing and data collection). This is also commensurate with our transfer findings, i.e., that hypothesizing and data collection transfer across content areas and over time.

4 Conclusions, Future Directions, and Implications

In this study we investigated the robustness of our scaffolding using students' performances on various inquiry practices at different time intervals and across different topics, thereby addressing far transfer. Our result showed, in general, that our scaffolding was robust for hypothesizing and collecting data practices because students' competencies continued to improve when evaluated after 40 days, 80 days, and 170 days regardless of topic difficulty. Specifically, we were interested in whether adaptive scaffolding of inquiry practices on one topic was enough to support student performance on different topics completed at various time intervals. As such, these represent metrics of far transfer. Despite the difficulty of moving from less difficult topics (i.e. animal and plant cell) to more advanced topics (i.e., genetics and natural selection), we found that the effects of scaffolding were still highly robust in terms of student performance relative to the first inquiry topic. This pattern was consistent across the practices of hypothesizing and collecting data. For the practices of interpreting data and warranting claims, growth was influenced by topic difficulty, as identified in prior studies [29].

In sum, these findings suggest that adaptive scaffolding in one topic in an ITS can benefit student inquiry learning even after scaffolding is removed and that the effect of adaptive scaffolding is maintained after long periods of time ranging from about 40 days to about 170 days. Our interpretation of these findings is that the procedural support given by Rex for inquiry practices is greatly supporting the acquisition and refinement of competencies, which undergirds students' inquiry. The effects of adaptive scaffolding were also apparent across topics in Life Science, some of which were more difficult than others. In future studies, we will take into account the difficulty of activities prior to analyses. We note that the success of far transfer of inquiry learning may depend on both increasing difficulty of inquiry practices and increasing complexity of inquiry topics. Overall, the findings in the present study inform assessment designers and researchers that, if properly designed, scaffolding aimed at supporting students' competencies at various inquiry practices can greatly benefit students' deep learning of and performance on inquiry practices such that their learning is robust and can be transferred to other topics, even when these topics are presented long after the original scaffolding. In future work, we will include a control group to take into account other activities in the classroom that occurred between opportunities to use the Inq-ITS system that may have effected inquiry transfer and include a separate pre-test measure outside of Inq-ITS. Additionally, it will be valuable to counterbalance the order of microworld activities. We are currently in the process of conducting a finer-grained analysis that will enable us to better understand how different factors (i.e. topic difficulty) interact and influence student performance on each inquiry practice.

References

1. Next Generation Science Standards Lead States: Next generation science standards: for states, by states. National Academies Press, Washington (2013)
2. Hmelo-Silver, C.E., Duncan, R.G., Chinn, C.A.: Scaffolding and achievement in problem-based and inquiry learning: a response to Kirschner, Sweller, and Clark (2006). Educ. Psychol. **42**, 99–107 (2007)
3. Kang, H., Thompson, J., Windschitl, M.: Creating opportunities for students to show what they know: the role of scaffolding in assessment tasks. Sci. Educ. **98**, 674–704 (2014)
4. McNeill, K.L., Krajcik, J.S.: Supporting grade 5-8 students in constructing explanations in science: the claim, evidence, and reasoning framework for talk and writing. Pearson (2011)
5. Quintana, C., Reiser, B.J., Davis, E.A., Krajcik, J., Fretz, E., Duncan, R.G., Kyza, E., Edelson, D., Soloway, E.: A scaffolding design framework for software to support science inquiry. J. Learn. Sci. **13**, 337–386 (2004)
6. Vygotsky, L.S.: Mind in society: the development of higher psychological processes. Harvard University Press, Cambridge (1978)
7. Staer, H., Goodrum, D., Hackling, M.: High school laboratory work in Western Australia: openness to inquiry. Res. Sci. Educ. **28**, 219–228 (1998)
8. Campbell, T., Zhang, D., Neilson, D.: Model based inquiry in the high school physics classroom: an exploratory study of implementation and outcomes. J. Sci. Educ. Technol. **20**, 258–269 (2011)
9. Deters, K.M.: Student opinions regarding inquiry-based labs. J. Chem. Educ. **82**, 1178–1180 (2005)
10. Brown, A.L., Campione, J.C.: Guided Discovery in a Community of Learners. The MIT Press, Cambridge (1994)
11. Noroozi, O., Kirschner, P.A., Biemans, H.J., Mulder, M.: Promoting argumentation competence: extending from first-to second-order scaffolding through adaptive fading. Educ. Psychol. Rev., 1–24 (2017)
12. Martin, N.D., Tissenbaum, C.D., Gnesdilow, D., Puntambekar, S.: Fading distributed scaffolds: the importance of complementarity between teacher and material scaffolds. Instr. Sci., 1–30 (2018)
13. McNeill, K.L., Lizotte, D.J., Krajcik, J., Marx, R.W.: Supporting students' construction of scientific explanations by fading scaffolds in instructional materials. J. Learn. Sci. **15**, 153–191 (2006)
14. Gobert, J.D., Moussavi, R., Li, H., Sao Pedro, M., Dickler, R.: Real-time scaffolding of students' online data interpretation during inquiry with Inq-ITS using educational data mining. In: Auer, M.E., Azad, A.K.M., Edwards, A., de Jong, T. (eds.) Cyber-Physical Laboratories in Engineering and Science Education, pp. 191–217. Springer, Cham (2018). https://doi.org/10.1007/978-3-319-76935-6_8
15. Quellmalz, E.S., Timms, M.J., Silberglitt, M.D., Buckley, B.C.: Science assessments for all: integrating science simulations into balanced state science assessment systems. J. Res. Sci. Teach. **49**, 363–393 (2012)
16. Tabak, I., Reiser, B.J.: Software-realized inquiry support for cultivating a disciplinary stance. Pragmat. Cogn. **16**, 307–355 (2008)
17. van Joolingen, W.R., de Jong, T., Lazonder, A.W., Savelsbergh, E.R., Manlove, S.: Co-Lab: research and development of an online learning environment for collaborative scientific discovery learning. Comput. Hum. Behav. **21**, 671–688 (2005)
18. Reiser, B.J.: Scaffolding complex learning: the mechanisms of structuring and problematizing student work. J. Learn. Sci. **13**, 273–304 (2004)

19. Klahr, D., Nigam, M.: The equivalence of learning paths in early science instruction: effects of direct instruction and discovery learning. Psychol. Sci. **15**, 661–667 (2004)
20. Koedinger, K.R., Anderson, J.R.: Illustrating principled design: the early evolution of a cognitive tutor for algebra symbolization. Interact. Learn. Environ. **5**, 161–180 (1998)
21. Gobert, J.D., Sao Pedro, M., Raziuddin, J., Baker, R.S.: From log files to assessment metrics: measuring students' science inquiry skills using educational data mining. J. Learn. Sci. **22**, 521–563 (2013)
22. Gobert, J.D., Sao Pedro, M.A., Baker, R.S., Toto, E., Montalvo, O.: Leveraging educational data mining for real-time performance assessment of scientific inquiry skills within microworlds. J. Educ. Data Min. **4**, 111–143 (2012)
23. Sao Pedro, M., Baker, R., Gobert, J.: Incorporating scaffolding and tutor context into bayesian knowledge tracing to predict inquiry skill acquisition. In: Proceedings of the 6th International Conference on Educational Data Mining, pp. 185–192. EDM Society (2013)
24. Sao Pedro, M.: Real-time assessment, prediction, and scaffolding of middle school students' data collection skills within physical science simulations. Worcester Polytechnic Institute, Worcester (2013)
25. Moussavi, R., Gobert, J., Sao Pedro, M.: The effect of scaffolding on the immediate transfer of students' data interpretation skills within science topics. In: Proceedings of the 12th International Conference of the Learning Sciences, pp. 1002–1005. Scopus, Ipswich (2016)
26. Moussavi, R.: Design, development, and evaluation of scaffolds for data interpretation practices during inquiry. Worcester Polytechnic Institute, Worcester (2018)
27. Li, H., Gobert, J., Dickler, R.: Automated assessment for scientific explanations in on-line science inquiry. In: Hu, X., Barnes, T., Hershkovitz, A., Paquette, L. (eds.) Proceedings of the 10th International Conference on Educational Data Mining, pp. 214–219. EDM Society, Wuhan (2017)
28. Corbett, A.T., Anderson, J. R., O'Brien, A.T.: Student modeling in the ACT programming tutor. Cogn. Diagn. Assess., 19–41 (1995)
29. Li, H., Gobert, J., Dickler, R., Moussavi, R.: The impact of multiple real-time scaffolding experiences on science inquiry practices. In: Nkambou, R., Azevedo, R., Vassileva, J. (eds.) ITS 2018. LNCS, vol. 10858, pp. 99–109. Springer, Cham (2018). https://doi.org/10.1007/978-3-319-91464-0_10
30. Cavallo, A.M.: Meaningful learning, reasoning ability, and students' understanding and problem solving of topics in genetics. J. Res. Sci. Teach. **33**, 625–656 (1996)
31. Corbett, A., Kauffman, L., Maclaren, B., Wagner, A., Jones, E.: A cognitive tutor for genetics problem solving: learning gains and student modeling. J. Educ. Comput. Res. **42**, 219–239 (2010)

Toward Real-Time System Adaptation Using Excitement Detection from Eye Tracking

Hamdi Ben Abdessalem(✉), Maher Chaouachi, Marwa Boukadida, and Claude Frasson

Department of Computer Science and Operations Research, University of Montreal, Montreal H3C 3J7, Canada
{benabdeh, chaouacm, boukadim, frasson}@iro.umontreal.ca

Abstract. Users' performance is known to be impacted by their emotional states. To better understand this relationship, different situations could be simulated during which the users' emotional reactions are analyzed through sensors like eye tracking and EEG. In addition, virtual reality environments provide an immersive simulation context that induces high intensity emotions such as excitement. Extracting excitement from EEG provides more precise measures then other methods, however it is not always possible to use EEG headset in virtual reality environment. In this paper we present an alternative approach to the use of EEG for excitement detection using only eye movements. Results showed that there is a correlation between eye movements and excitement index extracted from EEG. Five machine learning algorithms were used in order to predict excitement trend exclusively from eye tracking. Results revealed that we can detect the offline excitements trend directly from eye movements with a precision of 92% using deep neural network.

Keywords: Eye tracking · EEG · Excitement · Real-time adaptation · Artificial intelligence · Virtual reality · Emotional intelligence

1 Introduction

Virtual reality (VR) has shown a significant impact on the users' experience as it creates an immersive and individualized environment that allows a wider range of situations and interactions compared to other simulation techniques. Nevertheless, embedding VR environments with the ability to infer the users' affective states in real-time has been also an important topic, as these VR systems limit considerably the use of classical affective computing techniques. In fact, compared to traditional systems, VR headset covers the user s' face which makes techniques, such as facial emotion recognition and external judge methods impractical. Thus, these constraints hinder this VR technology to have access to affective data channels that could be used to adapt and optimize the environment to the user needs, particularly if VR is a serious game.

Serious game is a game designed for a primary purpose like education, training, simulation, exploring, analyzing, etc. rather than pure entertainment [1]. They could be used in many fields like education, medicine, military, etc. Serious games are generally known to induce a wide range of affective states on the users among which the

A. Coy et al. (Eds.): ITS 2019, LNCS 11528, pp. 214–223, 2019.
https://doi.org/10.1007/978-3-030-22244-4_26

excitement [2]. This last state is defined as the anticipation of a positively appraised energy-based event [3]. The level of excitement could be then an important indicator of the users' immersive experience as well as a quantitative metric of the degree of achievement of the game's objectives.

In addition, the development of artificial intelligence techniques and machine learning algorithms has significantly contributed to the integration of predictive models able to extract continuous information from multimodal data sources including sensors, cameras, context data, etc. [4, 5]. In this work, we propose a novel approach in analyzing the users' affective states using two sensing tools, electro-encephalography (EEG) and eye tracking integrated in a VR headset. Eye tracking is a method used to extract and analyze the users' eye movements. EEG is a technique to record brain activity and to infer mental states. We propose to use EEG measures to extract real-time information about users' excitement and explore how these measures are correlated with eye-tracking metrics. Then, we aim to build a real-time model able to predict excitement trends only from eye movements. To sum up, the two main hypotheses stated in this paper, are namely; **H1: are eye movements correlated to excitement?** And **H2: can we predict excitement trends from eye movements?**

This paper is organized as follows. In Sect. 2, we give an overview of the related works. In Sect. 3 we describe our approach and the physiological sensors that we use. In Sect. 4, we detail the experimental procedure, and finally in Sect. 5 we present the obtained results.

2 Related Works

2.1 EEG and Eye Tracking

EEG signals and eye tracking are often used in the fields of brain assessment and emotions detection. Using EEG signals, several researchers aimed to detect users' emotions for improving learning. Chaouachi et al. proposed an approach that extracts two mental state indexes from EEG which are engagement and workload using their system called "Mentor" [6]. They used some rules able to maintain students in a positive mental state while learning [7].

Moreover, Horlings and his colleagues [8] used EEG signals to recognize and classify the user's emotions and mental states. In fact, using EEG, they measured users' mental activity while they were exposed to different images. For that, they used the IAPS (International Affective Picture System) which contains images that provoke specific emotions. Results showed that EEG allows the recognition and classification of users' emotions.

2.2 Virtual Reality

Over the last few years, VR started to be used in many fields due to its remarkable advantages and the major one is immersion. In fact, VR tricks the mind of the user and increases his sense of presence in the virtual environment. It makes him believe that he is in a real world and promotes his performance [9]. Therefore, VR is being increasingly seen as the most interesting way to present an environment to the users.

Pedraza-Hueso et al. [10] introduced a VR system which consists of different types of exercises with which the user can train and rehabilitate several aspects such as cognitive capacities. Their system allows users to carry out physical and cognitive rehabilitation therapies using an interface based on Microsoft© Kinect.

Moreover Ghali et al. [11] designed a VR game to teach basic physics rules. In order to improve users' intuitive reasoning, they changed strategies of assistance in real-time according to player's levels of engagement and frustration. They noticed that VR offered an environment in which the user can deploy intuitive reasoning and acquire knowledge faster compared to usual academic training.

2.3 Artificial Intelligence and Emotions

Recently, emotions and human behavior have attracted the interest of researchers in computer science. Over the last few years, machine learning has gained more attention to solve many problems including emotion recognition and classification, which can be done either through text, speech, facial expression, gesture or EEG signals. Ang et al. [12] used decision trees as classifiers in order to detect annoyance and frustration based on prosody in human-computer dialogs. Results show that their prosodic model can predict whether the user is neutral, annoyed or frustrated with an accuracy identical to the agreement between human interlocutors.

Based on EEG signals, Chakladar and Chakraborty [13] have proposed a classification model that used Linear discriminant analysis (LDA) in order to classify four types of emotions (positive, negative, angry and harmony). The average accuracy of their model is 82%. Whereas Bhardwaj et al. [14], proposed an approach to detect and recognize seven emotions using Support-vector Machine (SVM) and LDA with an average overall accuracy of 74.13% and 66.50% respectively.

Based on facial expressions, Pitaloka et al. [15] proposed a classification of six basic emotions (angry, happy, disgust, fear, sad, and surprise) using an enhanced Convolution Neural Network (CNN) method. Moreover, Lopes et al. [16] developed facial expression recognition models with CNN and they achieved competitive results with 96.75% of accuracy.

In this research, we will use EEG in order to measure excitement, and eye tracking for analyzing eye movements and their relation with excitement. We will use VR in order to isolate the user and thus get more efficient measures due to its immersive effect, and finally we will use Machine Learning algorithms in order to correlate EEG and eye-tracking data.

3 Proposed Approach

In order to establish the correlation between eye movements and excitement, we propose to develop an adaptive system able to induce users' excitement.

3.1 Adaptive System

This system is composed of three main parts; the first one is the adaptable application (in this work, AmbuRun: a VR adaptable serious game), the second one is a measuring module and the third one is the neural agent.

AmbuRun

We started by creating AmbuRun, a serious VR game able to test our approach. This serious game should be adaptable according to users' excitement. AmbuRun consists of an ambulance carrying a sick person. The user takes control of the ambulance and tries to arrive safely at the hospital without damage in order to save the sick person. The user should dodge cars, buses, and trucks on the road to reach the hospital without harm [17]. The difference between cars and buses/trucks in the game play is that, if the user hits a bus or a truck (a big obstacle), the sick person will instantly die and the user must try again. However, if a car is hit, the health of the sick person will just decrease and he will not die instantly but only after multiple car hits.

We created this serious game in a way to support dynamic modifications using a neural agent which is described below. The possible modifications of AmbuRun concern two parameters: the speed and the difficulty. We vary the difficulty of the game by increasing and decreasing the obstacles' frequency. So, if the user encounters few cars and buses, the difficulty is easy, and if he cruises too many cars, the difficulty is hard. We change the speed of the game by increasing and decreasing the speed of the ambulance. We assume that the modification of these parameters affects the users' excitement, and thus, we can adapt this serious game according to it.

Measuring Module

The measuring module collects raw data from EEG and eye-tracking devices. It synchronizes all the received data, processes them, and then extracts indexes of emotions. In this work, we focus on the excitement as an output measure from the measuring module.

Neural Agent

The neural agent [18] is an intelligent agent designed to perform two main functions. The first one consists of receiving information about the virtual environment parameters and about the user's emotional state from the measuring module. The agent analyzes this information and decides of the best intervention/modification to be performed on the adaptable application in order to reach a desired emotional level. The second function aims to check that the resulting emotional state corresponds to what was expected, otherwise another intervention is triggered.

The neural agent runs in real time to analyze evolution of user's emotional state. It operates in a neurofeedback loop with the measuring module in order to change the emotional state of the user, which will indirectly trigger a modification of the virtual environment and adapt it to the user.

3.2 Measuring Sensors

In below section, we describe the eye-tracking and EEG systems that provide data to the measuring module.

Eye Tracking in VR

Our application is a VR application, so we chose the FOVE VR headset which has a built-in eye-tracking module. The device uses a 5.7-inch display with a WQHD (2560 × 1440) resolution, 100 degrees as a field of view and 70 fps (frame per seconds) frame rate. The eye-tracking module is composed of 2 infrared eye tracking systems (one for each eye) and has 120 fps frame rate with a tracking accuracy less than 1 degree. Fove VR headset provides a software in which we can monitor the movements of eyes in real-time. Figure 1 illustrates a screen capture of Fove software interface.

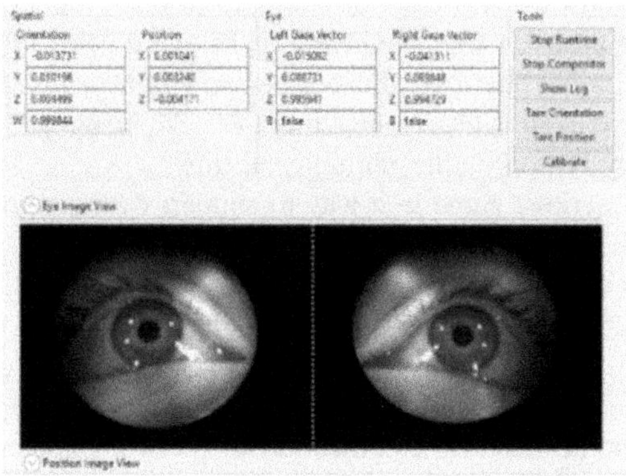

Fig. 1. Screen capture of Fove interface

Since Fove software output only provides eyes position in the three-dimensional space, a post-processing algorithm was developed in order to compute more meaningful metrics such as the eyes distance and the fixation period. We used the equation below to calculate the sum of the three-dimensional Euclidean space distance between two eye-tracking positions over a T time period.

$$\sum_{k=2}^{T} \sqrt{(x_k - x_{k-1})^2 + (y_k - y_{k-1})^2 + (z_k - z_{k-1})^2} \quad (1)$$

3.3 EEG Measures

We used Emotiv Epoc + EEG headset technology to track the excitement of the user. The headset contains 14 electrodes spatially organized according to international 10–20 system, moist with a saline solution. Emotiv system generates raw EEG data in (μV) with 128 Hz sampling rate as well as the five well-known frequency bands, namely Theta (4 to 8 Hz) Alpha (8 to 12 Hz), low Beta (12 to 16 Hz), high Beta (16 to 25 Hz) and gamma (25 to 45 Hz). Furthermore, the system uses internal algorithms to measure the following mental states: meditation, frustration, engagement, excitement and valence. Even though we don't have access to the system proprietary algorithms to infer these mental states from the raw data and the frequency bands, a number of studies have established the reliability of the output [19].

4 Experiments

In order to test our approach, we experimented our system on 20 participants (10 males and 10 females), with a mean age of 31.05 (SD = 4.96). Before taking part of the experiment, each user signed a consent form in which the goal of the study and the different steps of the experiment were clearly explained. Then, the user was equipped with the Emotiv Epoc+ and Fove VR headset devices described in the previous section. Once the user feels comfortable with the setup and ready to start, the measuring module, the neural agent as well as the AmbuRun VR game were simultaneously launched and the user starts interacting with the game using a wireless gamepad. Earphones connected to the device were used in order to isolate the player from the ambient environment and to intensify his level of immersion in the VR game. Figure 2 illustrates the experimental process.

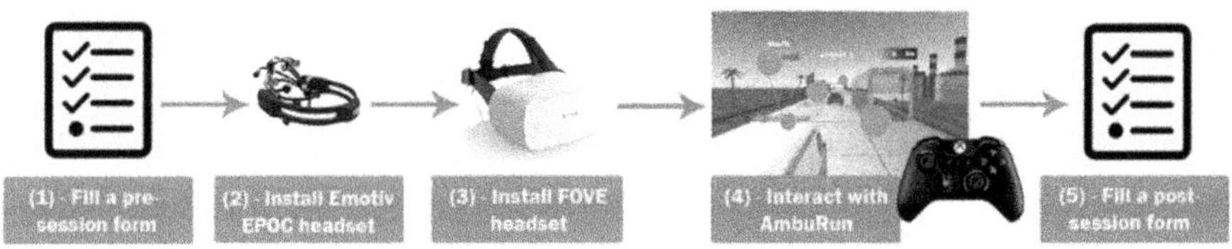

Fig. 2. Process of the experiment

During the experiment, we extracted and recorded two data sources: EEG data and eye movements data. The measuring module continuously tracks the excitement values of the user. The neural agent computes at a periodic time interval of 20 s the mean level of excitement and adapts accordingly the speed of the game. After finishing the experiment, each participant was asked to fill in a post-session questionnaire in which he provided his subjective feedback about the whole experience. The goal of this questionnaire was to help us improve our future research methodology.

5 Results and Discussion

In order to achieve our first goal and discover if there is a relationship between eye movements and excitement, we analyzed the participants' eye movements and excitement when the agent modifies the game parameters (speed and difficulty). Results showed that there is an impact of the agent modification on the excitement measured by EEG data. We noticed that there is a positive increase trend in the average of excitement 20 s after the agent raises the pace of the game. A repeated measure ANOVA with participants' excitement level as dependent variable revealed a significant increase of the agent intervention ($p = 0.000168$ and $F = 14.660$). Table 1 details the results and shows that when the agent makes the game faster, the mean excitement increases from 0.437 to 0.489 (5.2% more).

Table 1. ANOVA EEG excitement (more detailed study of this result could be found in [18])

	Excitement before	Excitement after
Mean	0.437	0.489
SD	0.214	0.203
N	220	200
F	14.660	
P	0.0001	

Likewise, we conducted a repeated measure ANOVA to analyze the relationship between eye movements mean distance before and after the agent's modification. Statistical results confirmed the existence of such a relationship with significant increase of the mean distance before and after this intervention ($p = 0.004$ and $F = 8.23732$). Table 2 details the obtained results.

Table 2. ANOVA Eye distance

	Eye distance before	Eye distance after
Mean	10.5216	12.2227
SD	5.9772	6.4467
N	220	200
F	8.23732	
P	0.004	

This result highlights the impact of the agent's intervention on the excitement measured by EEG and eye movements as well. In order to have a more fine-grained analysis at the intervention level (N = 220 interventions in total across all the participants), a Pearson correlation test between the eye movements difference and the mean excitement (i.e. before and after the agent intervention to increase the speed) was conducted. The results showed a significant fair correlation of 0.58 between these two measures ($p < 0.0001$). Hence, if the EEG excitement increases, the eyes move more which lead to our second research question: **can we predict excitement only from eye movements?**

In order to answer this question, we used our collected data which contains EEG excitement and eye tracking, then, we processed them by adding excitement "trend" column in which we set 0 if the EEG excitement decreases and 1 if it increases. Next, we deleted all EEG measures and we kept only the excitement trend column and eye-tracking data as our dataset. Then, we split our dataset randomly into 70% for training and 30% for testing. We trained the model in order to predict the excitement trend and we compared the predicted results on testing data with the real labels to analyze the accuracy of the algorithm. Five (5) supervised learning algorithms have been tested in our study, namely: Decision tree, Random forest, Support Vector Machine (SVM), Grid_search SVM and Deep Neural Network (DNN). For Decision tree, we used the Classification and Regression Trees (CART) version [20]. The random forest was set up with 200 estimators. For the SVM, C was set to 1.0 and gamma as default. Grid search SVM was also tested in order to find the best parameters for the SVM (best C = 10 and gamma = 0.0001). Finally, the DNN architecture used 4 hidden layers, with the first one composed of 10 neurons, and 20 neurons for the other 3 layers. Table 3 details the results of each algorithm.

Table 3. Classification reports of tested algorithms

		Precision	Recall	F1-score	Support
Decision tree	0	0.79	0.68	0.73	28
	1	0.79	0.87	0.82	38
	Avg/total	0.79	0.79	0.79	66
Random forest	0	0.81	0.79	0.80	28
	1	0.85	0.87	0.86	38
	Avg/total	0.83	0.83	0.83	66
SVM	0	0.79	0.68	0.73	28
	1	0.79	0.87	0.82	38
	Avg/total	0.79	0.79	0.79	66
Grid_search SVM	0	0.96	0.79	0.86	28
	1	0.86	0.97	0.91	38
	Avg/total	0.90	0.89	0.89	66
DNN	0	0.93	0.89	0.91	28
	1	0.92	0.95	0.94	38
	Avg/total	0.92	0.92	0.92	66

Average precision ranged from 79% from Decision tree and SVM to 92% from DNN. Grid_search SVM showed the highest precision for the 0 class (excitement decrease) with 96% and best recall for the 1 class (excitement increase). However, overall the DNN showed the best average precision, recall and f1-score.

The random baseline classifier which assigns the majority class (i.e. 1 in our case) gets an accuracy of 57% (38/66). Machine learning improved the accuracy by at least 22% (Decision tree) and at most 35% (DNN). This result underlines the impact of using machine learning with eye tracking data to detect users' excitement trend.

Table 4 shows the confusion matrix for each tested algorithm. For the 66 instances in the testing set, the DNN only misclassified 2 out of 38 instances in class 1 (increase of excitement trend) and 3 out of 28 instances in class 0 (decrease of excitement).

Table 4. Confusion matrix of tested algorithms

		Predicted 0	Predicted 1
Decision tree	Actual: 0	19	9
	Actual: 1	5	33
Random forest	Actual: 0	22	6
	Actual: 1	5	33
SVM	Actual: 0	19	9
	Actual: 1	5	33
Grid_search SVM	Actual: 0	22	6
	Actual: 1	1	37
DNN	Actual: 0	25	3
	Actual: 1	2	36

6 Conclusion

In this paper, we presented an approach which uses eye tracking as indicator of users' excitement level. Experiments were conducted during which the participants interacted with a VR serious game designed to be adaptable according to their excitement level measured using EEG signals. Results showed that there exists a significant correlation between the EEG excitement data and users' eye movement patterns. Five machine learning algorithms were tested in order to model excitement trend exclusively from eye-tracking data. Testing phase showed that, using DNN, we can predict the excitement trend from eye tracking with a precision of 92%. These results established that we can use eye movements in order to model users' excitement trend and consequently use it as a metric to intelligently adapt systems.

Acknowledgment. We acknowledge NSERC-CRD and Beam Me Up for funding this work.

References

1. Michael, D.: Serious Games: Games that Educate, Train and Inform. Thomson Course Technology, Boston (2006)
2. Schonauer, C., Pintaric, T., Kaufmann, H., Jansen - Kosterink, S., Vollenbroek-Hutten, M.: Chronic pain rehabilitation with a serious game using multimodal input. In: 2011 International Conference on Virtual Rehabilitation, pp. 1–8. IEEE, Zurich (2011)
3. Ganjoo, A.: Designing emotion-capable robots, one emotion at a time. In: Proceedings of the Annual Meeting of the Cognitive Science Society (2005)
4. Benlamine, M.S, Chaouachi, M., Frasson, C., Dufresne, A.: Predicting spontaneous facial expressions from EEG. Intell. Tutoring Syst. (2016)
5. Jraidi, I., Chaouachi, M., Frasson, C.: A hierarchical probabilistic framework for recognizing learners' interaction experience trends and emotions. Adv. Hum.-Comput. Interact. 1–16 (2014)

6. Chaouachi, M., Jraidi, I., Frasson, C.: MENTOR: a physiologically controlled tutoring system. In: Ricci, F., Bontcheva, K., Conlan, O., Lawless, S. (eds.) User Modeling, Adaptation and Personalization, pp. 56–67. Springer, Cham (2015). https://doi.org/10.1007/978-3-319-20267-9_5
7. Chaouachi, M., Jraidi, I., Frasson, C.: Adapting to learners' mental states using a physiological computing approach. In: Proceedings of the Twenty-Eighth International Florida Artificial Intelligence Research Society Conference, FLAIRS 2015, Hollywood, Florida, USA, 18–20 May 2015, pp. 257–262 (2015)
8. Horlings, R., Datcu, D., Rothkrantz, L.J.M.: Emotion recognition using brain activity. Presented at the (2008)
9. Biocca, F.: The Cyborg's dilemma: progressive embodiment in virtual environments. J. Comput.-Mediat. Commun. **3**, JCMC324 (2006)
10. Pedraza-Hueso, M., Martín-Calzón, S., Díaz-Pernas, F.J., Martínez-Zarzuela, M.: Rehabilitation using kinect-based games and virtual reality. Procedia Comput. Sci. **75**, 161–168 (2015)
11. Ghali, R., Abdessalem, H.B., Frasson C.: Improving intuitive reasoning through assistance strategies in a virtual reality game (2017)
12. Ang, J., Dhillon, R., Krupski, A., Shriberg, E., Stolcke, A.: Prosody-based automatic detection of annoyance and frustration in human-computer dialog. In: Hansen, J.H.L., Pellom, B.L. (eds.) INTERSPEECH. ISCA (2002)
13. Chakladar, D.D., Chakraborty, S.: EEG based emotion classification using "Correlation Based Subset Selection". Biol. Inspired Cogn. Archit. **24**, 98–106 (2018)
14. Bhardwaj, A., Gupta, A., Jain, P., Rani, A., Yadav, J.: Classification of human emotions from EEG signals using SVM and LDA classifiers. In: 2015 2nd International Conference on Signal Processing and Integrated Networks (SPIN), pp. 180–185. IEEE, Noida (2015)
15. Pitaloka, D.A., Wulandari, A., Basaruddin, T., Liliana, D.Y.: Enhancing CNN with preprocessing stage in automatic emotion recognition. Proc. Comput. Sci. **116**, 523–529 (2017)
16. Lopes, A.T., de Aguiar, E., De Souza, A.F., Oliveira-Santos, T.: Facial expression recognition with convolutional neural networks: coping with few data and the training sample order. Pattern Recognit. **61**, 610–628 (2017)
17. Ben Abdessalem, H., Frasson, C.: Real-time brain assessment for adaptive virtual reality game: a neurofeedback approach. Brain Function Assessment in Learning. LNCS (LNAI), vol. 10512, pp. 133–143. Springer, Cham (2017). https://doi.org/10.1007/978-3-319-67615-9_12
18. Ben Abdessalem, H., Boukadida, M., Frasson, C.: Virtual reality game adaptation using neurofeedback. In: The Thirty-First International Flairs Conference (2018)
19. Aspinall, P., Mavros, P., Coyne, R., Roe, J.: The urban brain: analysing outdoor physical activity with mobile EEG. Br. J. Sports Med. **49**, 272–276 (2015)
20. Loh, W.-Y.: Classification and regression trees: Classification and regression trees. Wiley Interdiscip. Rev. Data Min. Knowl. Discov. **1**, 14–23 (2011)

Towards Predicting Attention and Workload During Math Problem Solving

Ange Tato(✉), Roger Nkambou, and Ramla Ghali

Université du Québec à Montréal, Montreal, Canada
angetato@gmail.com

Abstract. Gifted students are characterized by a low level of attention and workload. Thus, it is very important to detect the variation of these values in real time when children are solving problems. A low value of workload or attention could be an indicator that the child is gifted. In this paper, we conducted a preliminary study in order to detect when children have a low values of attention or workload. A sample of 17 pupils participated in this study by solving math problems in an environment called Netmath. The EEG signal data collected from the experiment was used to train a Long Short Term Memory network (LSTM) to predict two mental states (attention and workload) in real time, when solving math problem. First results show that it is possible to predict these values in real time and the accuracy of the prediction is slightly above the random model. This pilot research provide some insight to the hypothesis that we can predict those variables in real time, which might be useful to intelligent tutor and to detect gifted children.

Keywords: Electroencephalography (EEG) prediction · LSTM · Attention · Workload

1 Introduction

It has been shown that gifted students have low values of attention and cognitive workload when solving problems [7]. Hence, if we are able to predict the attention and the workload of a student in a real time setting during problem solving, we will be able to predict when he/she will have great chance to succeed, and also be able to predict if he/she is gifted. While existing studies aim at classifying mental states in real time, predicting them have not been explore yet. Therefore we conducted a pilot study where we have developed a LSTM [8] model for predicting student's attention and cognitive workload when solving problems. The goal is to reach a model that would be able to accurately predict what the mental sate of the learner would be at time $t + 1$ knowing what it was at time $t, t - 1, ..., t - T$ where T represents the timestep of the model. Such a model can be useful in an intelligent tutor for helping infer student's mental states and decide accordingly what and how to tutor at each moment.

A. Coy et al. (Eds.): ITS 2019, LNCS 11528, pp. 224–229, 2019.
https://doi.org/10.1007/978-3-030-22244-4_27

The paper is organized as follows. Section 2 presents related work. Section 3 describes the model, the experiment settings following by the results and discussions. Finally, Sect. 4 presents our conclusions.

2 Related Work

2.1 Gifted Students

Historically, there are many definitions and conceptualizations of gifted and talented students [6,10,14]. Some authors allege that the high intellectual potential is innate (genetically present) and others that it represents the result of training or development of abilities or capacities of the child. Intellectual assessment or intelligence quotient remains an important indicator of giftedness. Two essential models are used to define giftedness [12], the one of Renzulli [10] and the other of Sternberg [11]. According to Renzulli [10], there are two types of gifted students: the first type corresponds to those with high academic potential. The second type corresponds to those with high creative potential. It proposes three components of skills to characterize the behavior of gifted children (intelligence, creativity and implication). These components interact. Sternberg [12] described a model of five criteria (excellence in one area relative to other people, scarcity of the level reached against peers, potential to produce something, ability to demonstrate skills with a valid assessment, and relative value of the skill for society).

2.2 Capturing Mental States

The EEG signal is a voltage signal that can be measured on the surface of the scalp, arising from large areas of coordinated neural activity manifested as synchronization (groups of neurons firing at the same rate) [4]. EEG signal can measurably detect variation in the neural activity. Attention and cognitive workload are mental states that can be tracked by a EEG headset. Many studies have used those mental states to predict reading behavior in a tutor [4], to predict performance level during cognitive task [3], to predict math problem solving [1] or to monitor indexes of alertness, cognition, and memory [2]. Berka and her team [2] used indexes extracted from EEG in order to study the engagement and workload mental states, Chaouachi et al. [5] integrated these two mental states in their system, Mentor. This system used some rules in order to maintain students in a positive mental state while learning, and reacted each time on selecting the appropriate next activity to present to the learner. This initial results appear to suggest that EEG might be a valuable technology for directly assessing a student's level of cognitive effort. However, none of the previous works led to a model for predicting the future mental state of the learner. Yet such information could be of interest if we want to prevent undesirable mental states that might affect the learning process. The ideal predictive model should preserve the temporal and sequential nature of data making the LSTM the best potential candidate.

2.3 Using Long Short Term Memory (LSTM) for Predicting Attention and Workload

LSTM has became a core component of neural Natural Language Processing (NLP) [13]. The LSTM architecture [8] can learn long-term dependencies using a memory cell that is able to preserve states over long periods of time. While numerous LSTM variants have been proposed, here we used the version proposed by Zaremba [16]. It is a model able to extract sequence information from data. It is suitable for problems where the temporal prediction is important as predicting attention in real time based on past values. Given a sequence $x_1, ..., x_T$ where x_i represents the concatenation of data depicting the mental states at time $t = i$ s and $T = 20$ s (time-step of our model), the LSTM processes each x at a time, keeping track of cell and state vectors (c_t, h_t) which are computed as a function of x_t, c_{t-1} [9]. In the next section, we describe how the model has been implemented in this study.

3 Experiments

We conducted an experiment where we asked elementary school students (4th and 5th grades) to solve the selected tasks from NetMath [1] environment which is a web application to support learning mathematics for primary and secondary students (from 3rd primary grade to 4th secondary grade). The proposed tasks were designed for higher-level students (6th grade). 17 students (10 F, 7 M) voluntarily participated in this study. Students are aged between 9 and 11 years (M = 10.05; SD = 0.42). We choose a total of 10 tasks from the platform. These tasks are divided into three levels of difficulty: easy, medium and hard. During the fulfillment of the tasks, we collected real time data from Neeuro Senzeband non-invasive EEG headset. This headset allows us to obtain EEG raw data from 4 channels and three mental states measures (Attention, Workload and Relaxation).

3.1 Dataset

Each of the 17 participants spent around 30 min to solve all the problems. For each of these participants we have gathered their attention, cognitive workload, relaxation, both hemispheres of the brain left/right, center-left and center-right at each second. For the prediction of attention and workload, we have defined a step of 20 s which defined the timestep for the LSTM model. The dataset was thus divided into input data X and output data Y where each line of X represents events that happened 20 s before and each line of Y is a label which is the event that happened at the 21st second. This configuration gave us around 20000 data where 75% where used for training and 25% for test. Figure 1 shows an excerpt of the dataset.

[1] https://www.netmath.ca/fr-qc/.

Time	Attention	Workload	Relaxation	Left	CenterLeft	CenterRight	Right
16:13:29 PM	0,998691	0,3732375	0,1199541	-6	-13	-8	-2
16:13:30 PM	0,01	0,01	0,07488824	8	10	68	15
16:13:32 PM	0,89289	0,01	0,01517821	-14	11	113	29
16:13:32 PM	0,89692	0,01	0,05044472	-6	-7	6	10
16:13:33 PM	0,904556	-0,07081398	0,07155754	15	-6	-46	-5
16:13:34 PM	1	0,2673379	0,06456525	13	11	-9	9
16:13:35 PM	0,968494	0,01	0,0776386	8	4	-2	6
16:13:36 PM	0,969414	0,5121103	0,09333985	50	9	9	16
16:13:37 PM	0,359994	0,01	0,0469471	47	-2	-21	9
16:13:38 PM	0,724037	0,276055	0,09904281	39	4	3	13
16:13:39 PM	0,970849	0,333482	0,1598548	19	3	6	9

Fig. 1. Excerpt of the raw EEG data.

3.2 The Model

Initially we trained a regression model (LSTM with mean squared error) whose objective was to predict the approximate value of attention/workload but this first model gave us poor results. We have therefore turned the problem into a classification problem where we predict whether the value of the attention is greater than a threshold value or not. The best threshold value found is 0.6 (0.02 for the cognitive workload). So the designed model is able to predict whether attention will be high (>0.6) or low (≤0.6). For example, if the value to predict is less than 0.6, then the label is encoded as [1,0] else the label is encoded as [0,1]. The model (see Fig. 2) is composed of an input layer containing 7 neurons representing each of the 7 variables presented above. It turns out that when adding or removing both hemispheres of the brain left/right data as input, the results did not changed considerably. Therefore, the final model does not include those 4 attributes. The raw values were retained because the standardization and normalization processes of the data gave us poor results. After the input layer, there is the LSTM layer able to extract the sequential information useful for the prediction. The penultimate layer summarizes the information extracted in the lower layers in order to send it to the final layer that makes the final prediction. The activation function ReLU (Rectified Linear Unit) is used in all layers except for the last layer where the activation function is the Softmax. The loss is the binary cross entropy. The number of epoch was fixed to 50 and the batch size to 500. We used Adam as the optimizer. We implemented the models in python using Keras.

3.3 Results and Discussions: Can We Predict Attention and Workload?

The model that predicts the attention gave 61% of accuracy and the model that predict the workload gave 76% of accuracy which is the average on 20 different training. Table 1 gives examples of predicted values vs real values of the two models. The second model performs better than the first model. These results show that it is possible to observe the variation of the attention and the workload of students during problem solving and predict how those mental states will vary in real time as students continue to solve problems. However,

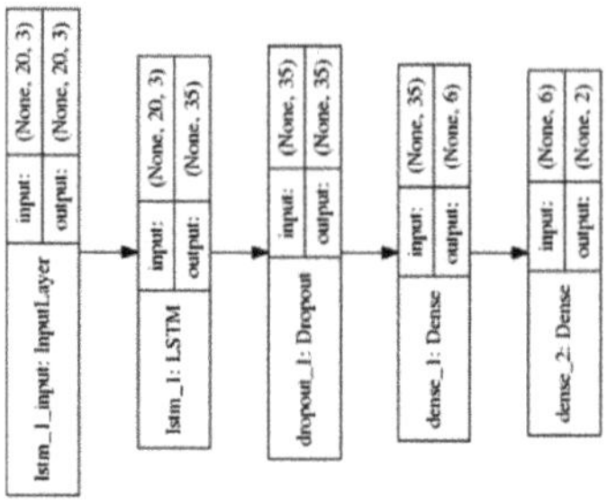

Fig. 2. Model to predict attention and workload in real time.

we noticed that the first model is slightly above the random model (better than chance) which means somehow that it is difficult to predict how the attention will vary across time compare to the workload. The model as designed is a general model in the sense that it does not take into account the specificities of each person. However, since people are different and may display different behaviours for the same task, it is important to incorporate a more contextual data to that model. One solution would be to add into these models a branch by using the attention mechanism [15] that would take as input specific attributes of the person whose attention or cognitive workload is to be predicted. The models need more contextual data to be able to figure out how the variables change across time. Nonetheless, our results seem promising, but are they meaningful? EEG-based mental state detectors will presumably need to be much more accurate in the first place to help the design of accurate predictor and to help intelligent tutors in real time.

Table 1. Examples of prediction vs real data. [0, 1] for Attention means that the value to predict is above 0.6 and [1, 0] means that the value to predict is less than 0.6. [0, 1] for Workload means that the value to predict is above 0.02.

Output	Real values	Predicted values
Attention	[0, 1]	[0.46971744, 0.5302825]
Attention	[1, 0]	[0.590336, 0.40963754]
Workload	[0, 1]	[0.495932, 0.504068]
Workload	[1, 0]	[0.9372175, 0.0627825]

4 Conclusion

This preliminary study suggests that we can predict mental states variation across time. We found weak but above chance performance for predicting Attention. Even if the models built do not gave us good results, we have shown that it is possible to predict mental state across time based on past events. However the models need more contextual data to be able to predict more accurately the variations over time. As an example, we could include specific information on the student such as age, or the type of exercise they are currently solving. Our next step will be to first cluster students based on their behaviours and then to feed that new information to the models. The resulting solution will then be tested in a real-time setting.

Acknowledgments. We would like to thanks The Fonds de recherche du Québec – Nature et technologies (FRQNT) for their financial support and Beam Me Up Games.

References

1. Beal, C.R., Galan, F.C.: EEG estimates of cognitive workload and engagement predict math problem solving outcomes. Soc. Res. Educ. Eff. Consultado el, 05/17/19 (2012). http://eric.ed.gov/?q=EEG&id=ED536318
2. Berka, C., et al.: Real-time analysis of EEG indexes of alertness, cognition, and memory acquired with a wireless EEG headset. Int. J. Hum.-Comput. Interact. **17**(2), 151–170 (2004)
3. Besserve, M., Philippe, M., Florence, G., Laurent, F., Garnero, L., Martinerie, J.: Prediction of performance level during a cognitive task from ongoing EEG oscillatory activities. Clin. Neurophysiol. **119**(4), 897–908 (2008)
4. Chang, K.M., Nelson, J., Pant, U., Mostow, J.: Toward exploiting EEG input in a reading tutor. Int. J. Artif. Intell. Educ. **22**(1–2), 19–38 (2013)
5. Chaouachi, M., Jraidi, I., Frasson, C.: Adapting to learners' mental states using a physiological computing approach. In: FLAIRS Conference, pp. 257–262 (2015)
6. Davis, G.A., Rimm, S.B.: Education of the Gifted and Talented. Prentice-Hall, Inc., Upper Saddle River (1989)
7. Ghali, R., Abdessalem, H.B., Frasson, C., Nkambou, R.: Identifying brain characteristics of bright students. J. Intell. Learn. Syst. Appl. **10**(03), 93 (2018)
8. Hochreiter, S., Schmidhuber, J.: Long short-term memory. Neural Comput. **9**(8), 1735–1780 (1997)
9. Murdoch, W.J., Szlam, A.: Automatic rule extraction from long short term memory networks. arXiv preprint arXiv:1702.02540 (2017)
10. Renzulli, J.S.: The Three-Ring Conception of Giftedness: A Developmental Model for Promoting Creative Productivity. Cambridge University Press, Cambridge (2005)
11. Sternberg, R.J.: Handbook of Intelligence. Cambridge University Press, Cambridge (2000)
12. Sternberg, R.J., Jarvin, L., Grigorenko, E.L.: Explorations in Giftedness. Cambridge University Press, Cambridge (2010)
13. Tai, K.S., Socher, R., Manning, C.D.: Improved semantic representations from tree-structured long short-term memory networks. arXiv preprint arXiv:1503.00075 (2015)
14. Terman, L.M., Baldwin, B.T., Bronson, E., De Voss, J.C.: Mental and Physical Traits of a Thousand Gifted Children, vol. 1. Stanford University Press, Stanford (1926)
15. Xu, K., et al.: Show, attend and tell: neural image caption generation with visual attention. In: International Conference on Machine Learning, pp. 2048–2057 (2015)
16. Zaremba, W., Sutskever, I.: Learning to execute. arXiv preprint arXiv:1410.4615 (2014)

Poster Papers

Agents' Cognitive vs. Socio-Affective Support in Response to Learner's Confusion

Zhou Long[1,2], Dehong Luo[2], Sheng Xu[1], and Xiangen Hu[1,3](✉)

[1] Central China Normal University, Wuhan 430079, China
xiangenhu@gmail.com
[2] Huaihua University, Huaihua 418000, China
[3] University of Memphis, Memphis 38152, USA

Abstract. Drawing on a social-regulating approach, we experimentally compared the effects of cognitive and socio-affective support on learning outcome in an intelligent tutoring system. Findings show that cognitive support from pedagogical agents in response to learner's confusion is preferable for enhancing students' learning outcomes.

Keywords: Confusion · Cognitive support · Socio-affective support

1 Introduction

Confusion represents the "wisdom of the emotions" [1], providing time-tested learning opportunity in Intelligent Tutoring Systems (ITS) environments. Specifically, *confusion* is an epistemic affective state [2], indicating that there is a problem with the current state of one's knowledge. Crucially, however, confusion is not always helpful when it remains unresolved. Some learners regulate confusion by themselves, and others may need additional support to get out of the sustained confusion state.

When we experience protracted confusion, we typically feel the urge to tell or ask others about our experience. This phenomenon is seen as social confusion regulation. Listeners may primarily offer *cognitive support*, which is directed at altering cognition related to the emotional experience, or *socio-affective support*, which includes comfort and validation [3]. It has been argued that these two types of support exert different influence on how a person thinks, feels, and acts [3]. Then, is it apply to learning domain? We aimed to compare the effects of cognitive and socio-affective support on confusion learning outcome in ITS environments. In this study, the ITS environments that were specifically designed to trigger and regulate confusion during research method learning have been developed.

2 Method and Results

Undergraduate participants. 84 undergraduates at a general university in China were recruited to participate in exchange for extra course credits, who had no learning experience in experimental material (research method). Three volunteers were dropped from the dataset because their finishing time of experiment was over 3 standard

A. Coy et al. (Eds.): ITS 2019, LNCS 11528, pp. 233–235, 2019.
https://doi.org/10.1007/978-3-030-22244-4

deviations above average time. This resulted in a final sample of 81 participants (54 female and 27 males, mean age = 21.2 yrs).

Mixed design. The study involved a 4 (Social Confusion Induction: true-false, false-true, false-false, true-true) x 3 (Social Confusion Regulation: cognitive support, socio-affective support, no support) mixed design. Participants have received all four types of social confusion induction in a Graeco-Latin Square order and randomly assigned to one of the social confusion regulation conditions. Learning outcome served as the dependent variable was the score gap between post-test and mid-test.

Social confusion induction manipulation. Similar to D'Mello et al. [4], social confusion induction was operationalized by varying contradictory information in agent agreement and information correctness during the trialogues (three-party conversation: a participant and two pedagogical peer agents) phase. In the control condition, both agents agreed on the correct information (true-true), while in the other three experimental conditions, two agents either disagreed with each other or agreed with the incorrect information. After both agents presented their respective opinions, then the participant would be asked by an agent to express oneself. The contradiction between agents' opinion was expected to trigger the participant's confusion.

Social confusion regulation manipulation. We operationalized social confusion regulation by support varied in types. *Cognitive support* was always characterized by triggering participants to stop, reflect, and further deliberate over which agent's opinion was correct and why that opinion was correct (e.g. "XX, remember to think about how students in the control and experimental groups behaved during the study. Try to put together a convincing argument to get me on your side."). *Socio-affective support* messages from agents always included validation of participants' confusion, understanding, and encouraging (e.g. "You know, this feeling is actually a good thing in learning. It helps us to notice that we ignore some knowledge about experimental groups. Let's keep trying to figure out this concept."). All supportive reactions were tailored to the specific trialogue background to enhance ecological validity.

Procedure. The experiment occurred over five phases (each for 2.5 h): the participants (1) took pretest for prior knowledge, (2) acquired research method knowledge through multimedia learning to identify the contradictory of information in later trialogues, (3) took mid-test to assess and control over learning outcome in multimedia learning, (4) attended eight trialogues (each about one concept) offering contradictory and supporting information to induce and regulate participant's confusion respectively, and (5) took post-test to check each one's overall learning outcome. Each trialogue in the fourth phase began with a description of a research method practice. Participants read the description and then discussed it with the agents. Each discussion involved four trials. The first three trials were about social confusion induction, and the last trial was for social confusion regulation.

Learning outcome measurement. Learning content about eight concepts of research method covered in eight trialogues was tested three times, including pretest, mid-test, and post-test. Learning outcome served as the dependent variable was used to assess the benefit of support, indicated by the score gap between post-test and mid-test. Each test had 24 multiple-choice questions with three questions per concept. The three types of items were based on the first three levels of Bloom's Taxonomy (knowledge, comprehension, application). Three alternate test versions and assignment were counterbalanced across participants.

Results of learning outcome. To test which type of support benefited learning outcome, and whether these effects were dependent on the confusion occurrence, a 4(Social Confusion Induction) x 3(Social Confusion Regulation) two-way ANOVA was performed with learning outcome as dependent measures. The proportional occurrence of test scores for learning outcome are presented in Table 1. The results showed a significant interaction between social confusion induction and social confusion regulation, $F(6, 234) = 2.46$, $p < .001$, $\eta_p^2 = .11$. Simple-effects analyses suggested that within the true-false condition, the participants who received cognitive support outperformed those who received socio-affective support ($M_{CS\text{-}SAS} = .22$, $SD = .06$, $p = .001$) and no support ($M_{CS\text{-}NS} = .25$, $SD = .06$, $p < .001$); similar learning benefits trend existed in the false-true condition ($M_{CS\text{-}SAS} = .2$, $SD = .07$, $p = .02$; $M_{CS\text{-}NS} = .24$, $SD = .06$, $p = .001$), but not in the true-true and false-false condition.

Table 1. Means (M) and Standard Deviations (SD) of learning outcomes.

	CS ($N = 27$) *M (SD)*	SAS ($N = 27$) *M (SD)*	NS ($N = 27$) *M (SD)*	Total ($N = 81$) *M (SD)*
True-False	.57 (.27)	.36 (.19)	.32 (.13)	.42 (.23)
False-True	.55 (.25)	.35 (.22)	.32 (.23)	.41 (.26)
False-False	.31 (.2)	.28 (.13)	.25 (.09)	.28 (.15)
True-True	.34 (.25)	.44 (.25)	.26 (.14)	.35 (.23)

Notes. CS = cognitive support, SAS = socio-affective support, NS = No support.

In sum, we have successfully regulated confusion mainly by cognitive support in ITS environments. It should be noted that we assessed cognitive and socio-affective support separately, while we usually received both supports in daily life. The next step is to investigate a combination of both, and whether the temporal order of received support might be relevant. The findings help to build ITS more effective.

Acknowledgments. We would like to thank the support from Jingshi Lexue Education Technology Co., and the Collaborative Innovation Project (Grant NO. 2016-04-011-BZK01).

References

1. Lazarus, R.S.: Emotion and adaptation. Oxford University Press, Oxford (1991)
2. Pekrun, R., Linnenbrink-Garcia, L.: Academic Emotions and Student Engagement. In: Christenson, S., Reschly, A., Wylie, C. (eds) Handbook of Research on Student Engagement, pp. 259–282. Springer, Boston (2012)
3. Rimé, B.: Emotion elicits the social sharing of emotion: theory and empirical review. Emot. Rev. **1**, 60–85 (2009)
4. D'Mello, S., Lehman, B., Pekrun, R., Graesser, A.: Confusion can be beneficial for learning. Learn. Instr. **29**, 153–170 (2014)

Patterns of Collaboration Dialogue Acts in Typed-Chat Group Problem-Solving

Duy Bui[1], Matthew Trotter[1], Jung Hee Kim[1], and Michael Glass[2(✉)]

[1] North Carolina A&T State University, Greensboro, USA
{dqbui,mtrotter}@aggies.ncat.edu, jungkim@ncat.edu
[2] Valparaiso University, Valparaiso, USA
michael.glass@valpo.edu

Abstract. This study observes patterns of dialogue acts in typed-chat student group problem-solving. Quantitative differences are observed according to the relative preparedness of the students within the discussion.

Keyword: Collaborative problem-solving dialogue

1 Introduction and Background

The COMPS project administers computer-mediated problem-solving chat dialogues in undergraduate college classes [1]. A goal of COMPS research is to characterize typical problem-solving dialogue activity. Characterizing typical activity advances toward a future goal of providing assessments of the chat activities, so the instructor can gauge whether students and chat groups are collaborating productively.

Our research hypothesis is that the students in a problem-solving dialogue behave differently according to their relative level of knowledge: the best-prepared student within the group may take on a different role than the least-prepared. To test this, we examined the frequencies of several categories of collaborative-activity dialogue acts.

This study used approximately a thousand dialogue turns of COMPS project transcripts manually tagged according to four categories of collaboration dialogue acts. Then we measured the different frequencies of the dialogue acts according to preparedness rank of the student.

The students in this study were in a 2nd-semester undergraduate computer programming class. They worked together in three-person groups for approximately an hour. The exercise in this study involved analyzing the object-oriented aspects of some Java code. The exercise prompt instructs the students to come to an agreement on successive segments of the problem. A teaching assistant (TA) then joins the discussion at the end of each segment to pass judgment and perhaps provide assistance. This script is intended to promote mutual dependence (the students don't signal for the TA until they all agree) and accountability (students should exhibit understanding for each part of the problem). An extract from one of these conversations is shown in Fig. 1.

Pre- and post-tests were administered before and after the dialogues. We ranked the three students in a discussion group, based on pre-test score. Rank 1 is the student within the discussion who was most prepared, rank 3 was the least prepared student.

A. Coy et al. (Eds.): ITS 2019, LNCS 11528, pp. 236–238, 2019.
https://doi.org/10.1007/978-3-030-22244-4

Person	Text	Acts	Sub-category
St1	public String toStrong(){ String result = null; result = lendingInstitution +' '+ PA-mount +' '+ iRate +' '+ etc.	A	Sharing Idea
St2	lol yall going in i think thats right tho	D, B	Joking, Agreement.
St1	we just have to explain the getters and setters now	C	Suggest next step
St3	Student 1 can u explain them	C	Check understanding
St1	besides excapsulation, accessors make it easier to change future things mybad on the spelling	A, D	Explanation, politeness
St4	So everything except the setters and getters are explained right?	C	Reflect
St1	encapsultion allows validation	A	Continue explanation
St3	I dont believe we've explained the properties	C	Suggest next step

Fig. 1. Extract of dialogue, with dialogue acts annotated.

We computed learning gains as (post–pre score)/(1–pre). We also measured each student's participation based on numbers of turns, using a formula which normalized according to the number of participants. Thus in a 3-person group, a person contributing 1/3 of the turns was recorded with a participation statistic of 0.5.

This study annotated the type of collaborative dialogue act in 1000 turns of dialogue from 10 group discussions. The four categories of dialogue acts are as follows. The categories are adopted from a project of one of the testing services which is building assessments of student collaborative conversations [2].

(A) Sharing ideas, or pointing at where ideas can be found
(B) Negotiating, including agreement/disagreement
(C) Regulating problem-solving
(D) Maintaining communication, e.g. politeness, joking or small talk

Dialogue acts do not exactly correspond to typed-chat dialogue turns, where the student presses <enter> to mark a turn. One dialogue turn can contain several acts, or a single person can utter a several-turns-long dialogue act without interruption by other students. Another complication is COMPS chat software permits overlapping dialogue turns, everybody can type at once and observe each other's typing without interruption [3]. In these transcripts overlapped turns are serialized according to when they ended. Note also that conversation segments with the TA present might plausibly show different student behaviors as they are no longer solving the problem but checking their answers. These segments were not counted in our results here.

2 Results

Table 1 shows how the differently prepared students had different conversational behaviors within a discussion. Table 1 records behaviors from the 30 students in the 10 discussion groups, all working on the same problem in the same semester. Most important is the relative mix of dialogue acts. Chi-squared tests show that the rank 3 least prepared students significantly differ in their mix of dialogue acts from rank 2 ($p < 0.05$) and from rank 1 ($p < 0.01$). A rank 3 student also participates significantly less than a rank 1, contributing fewer dialogue acts in each dialogue ($p < 0.01$). Consistent with previous COMPS results, the most prepared students showed little or no learning gain, while the least prepared showed the most [1].

Table 1. Different styles of contribution, based on relative preparedness within the group.

	Rank 1: n = 10	Rank 2: n = 10	Rank 3: n = 10
Avg learning gain	0.0	0.1	0.5
Numb. dialogue acts	442	311	220
A: sharing	30%	27%	25%
B: negotiating	28%	33%	33%
C: regulating	28%	27%	21%
D: maintaining	14%	13%	22%

A conclusion is: learning gains do not directly correlate with problem-solving participation acts. The lowest rank students had the largest learning gains, with lower percent of problem-solving dialogue acts in categories (A) through (C), and a higher proportion of category (D) acts which do not contain problem-solving content. Assessing collaborative dialogues may need to take this differential into account.

References

1. Kim, J.H., Kim, T., Glass, M.: Early experience with computer supported collaborative exercises for a 2nd semester java class. J. Comput. Sci. Coll. **32**(2), 68–86 (2016)
2. Hao, J., Liu, L., von Davier, A., Kyllonen, P.C.: Initial steps towards a standardized assessment for collaborative problem solving (CPS): practical challenges and strategies. In: von Davier, A., Zhu, M., Kyllonen, P.C. (eds.) Innovative Assessment of Collaboration. Methodology of Educational Measurement and Assessment, pp. 135–156, Springer (2017)
3. Glass, M., Kim, J.H., Bryant, K., Desjarlais, M.: Come let us chat together: simultaneous typed-chat in computer-supported collaborative dialogue. J. Comput. Sci. Coll. **31**(2), 96–105 (2015)

Analyzing Best Hints for a Programming ITS

Reva Freedman(✉) and Ben Kluga

Department of Computer Science, Northern Illinois University, Dekalb, USA
rfreedman@niu.edu, benkluga@gmail.com

Abstract. We are building an ITS that tutors students in beginning C++. It has the form of an intelligent review sheet that gives students increasingly difficult problems to solve, and gives them hints when needed. In this paper we evaluate proposed hints for this system. Our goal is to find the types of hints preferred by expert instructors.

Keywords: Intelligent tutoring systems · Computer science education · Teaching beginning programming

1 Introduction

We are developing a system that tutors students in beginning C++. It is an "intelligent review sheet" that gives students increasingly difficult problems to solve in categories they need to master, and provides hints when students give wrong answers. If a student gets a question wrong the second time, the system attempts to give a good followup hint. After that it provides the answer.

Due to the design of the student model, in the initial version of the system hints will be determined by the question rather than by the incorrect answer supplied by the student. We surveyed a set of expert instructors, all of whom had taught the course material for at least ten years, to see which types of hints they preferred both as initial hints and as followup hints.

In this paper we examine one categorization of hints to better understand the structure of preferred hints. This approach is derived from a line of research starting with Zhou et al. [2]. The categorization was developed to provide input to the algorithm we plan to use to generate the hints automatically.

Table 1 shows a sample problem to be used in the system. The problems are multiple choice with four possible answers and are presented via a web interface. The majority of problems ask the student to find the result of executing some C++ code. The remaining problems test the student's knowledge of C++ concepts, such as declarations, loops and classes. The system covers all of the topics in a one-semester beginning C++ class in the order that they are found in the course. In the analysis in this paper, we examine only the results from the code questions because hints for content questions are not in the scope of the algorithm mentioned above.

A. Coy et al. (Eds.): ITS 2019, LNCS 11528, pp. 239–241, 2019.
https://doi.org/10.1007/978-3-030-22244-4

Table 1. Sample question from C++ hint evaluation questionnaire.

Q2. If x is initialized to 3, what is the value of x after x *= 5 is executed?

a. 5. b. 15. c. 3. d. Cannot be determined from the information given.

____ Hint 1: What does the * operator do?
____ Hint 2: Compound assignment operators assign the result to the left operand.
____ Hint 3: What is the value of x after the instruction x=x*5 is executed?
____ Hint 4: What is x *= 5 an abbreviation for?
____ Hint 5: x *= 5 is an abbreviation for x = x*5.
____ Hint 6: *= is a compound assignment operator. It abbreviates x=x*5 to x*=5.
____ Hint 7: x*= 5 is an abbreviation. What do you think it is an abbreviation for?
____ Hint 8: Is *= an abbreviation? What do you think it could stand for?
____ Hint 9: (writein) ______________________________.

(i) Assume the student has given a wrong answer. Label the best hint with a 1.
(ii) Label the second best hint with a 2.
(iii) Go back to the original best hint. Suppose the student gets the question wrong again after getting that hint. Label the best followup hint with a 3.

2 Background: Hint Categorization

The problems requiring students to evaluate C++ code included hints in three main categories: rules, instantiated rules and pointers. A rule is a piece of content that the student needs to solve a problem. For example, in the problem above, "Compound assignment operators assign the result to the left operand" is a rule.

An instantiated rule is a version of a rule that has been instantiated with values that apply to the given problem. In the example above, "*= is a compound assignment operator. It abbreviates x = x*5 to x* = 5" is an instantiated rule where the operator, the left-hand operand and the right-hand operand have been instantiated to the values used in the problem.

A pointer is a hint that does not directly give the student content, but just refers to it. Most frequently, pointer hints are questions whose answer contains the desired content. The example above includes several pointer hints, including "What is x * = 5 an abbreviation for?". The term is based on the terminology used in [1].

These categories were chosen because they will be the basis of a planned logic engine that will automatically generate instantiated hints and pointers from rules. The survey also included two other categories of hints for code questions, analogies and remediation of misconceptions, that do not occur in the example above. Two final categories, mnemonics and C++ syntactic schemata, only occurred in the hints for conceptual questions and not in the hints for code questions.

3 Procedure

We prepared a questionnaire with 10 sample questions that will be used in the eventual system. For each question 8–10 hints were provided. Respondents were also given the opportunity to write in their own hints.

We asked the respondents to answer the three questions labeled (i)-(iii) in Table 1.

4 Results

The first line of Table 2 shows the results of the categorization. The totals include all hints that our expert instructors chose as potential responses to initial student errors, including both their first and second choices, which may have been writein hints. We did not include the followup hints, which were intended for a separate analysis on conversational coherence. We obtained the expected values by counting the total number of hints available in each category. These results give $\chi^2 = 22.12$ with $df = 4$, which is significant at the $p < .001$ level.

Table 2. Hints preferred for coding questions.

	Ptr	Rule	IRule	Miscon	Ana.
Observed values	28	17	9	7	2
Expected values	15	8	7	6	2

5 Conclusions and Future Work

Our results show that experienced instructors implicitly make distinctions in the types of hints they use.

In future work we plan to examine the information content of the followup hints to see whether information provided only in pointer form in the original hints is provided in more explicit form in the followup hints. We also plan to investigate whether individuals who prefer one type of hint as their first choice maintain that preference in followup hints.

References

1. Hume, G., Michael, J., Rovick, A., Evens, M.: Hinting as a tactic in one-on-one tutoring. J. Learn. Sci. **5**(1), 23–47 (1996)
2. Zhou, Y., Freedman, R., Glass, M., Michael, J., Rovick, A., Evens, M.: What should the tutor do when the student cannot answer a question? In: Proceedings of the 12th International FLAIRS Conference (1999)

Using a Simulator to Choose the Best Hints in a Reinforcement Learning-Based Multimodal ITS

Manohar Sai Jasti and Reva Freedman(✉)

Department of Computer Science, Northern Illinois University, Dekalb, USA
z1833049@students.niu.edu, rfreedman@niu.edu

Abstract. We are building a reinforcement learning-based multimodal ITS for human physiology. We used a panel of students to estimate the time it would take students to process the different types of hints given by our system. We used these data as the basis for a simulator to see which types of hints would be most worthwhile. Our system improves upon earlier systems by taking into account the differential cost of providing hints in three modalities.

Keywords: Intelligent tutoring systems · Reinforcement learning · Biology education

1 Introduction

We are building a multimodal ITS for teaching undergraduate human physiology. The system is based on a set of real-world scenarios presented to the student via a web interface. For each scenario given, the system will choose one or more questions to ask the student. A sample scenario and a question related to that scenario are shown in Table 1. The questions differ in difficulty level and in the type of answer requested. The system is based on a concept map that students are introduced to in lecture.

When a student gets a question wrong, the system provides a hint. We have categorized student errors based on the knowledge that the student needed to answer the question correctly. For each knowledge item, the system can reply with one of three types of hints.

- *Textual hint.* A one-sentence item shown on the screen, either a statement providing information or a question pointing the student at the desired content.
- *Visual hint.* An image displayed on the screen, e.g., a copy of the concept map, possibly with highlighting to focus the student's attention.
- *Video hint.* A short video clip, generally showing the course instructor explaining a topic or clarifying a misconception.

One of the goals of the project is to investigate which media are most useful for teaching physiology. Thus the three types of hints cover the same content using different media. The system contains a reinforcement learning module to identify which hints help students learn the material most efficiently.

A. Coy et al. (Eds.): ITS 2019, LNCS 11528, pp. 242–244, 2019.
https://doi.org/10.1007/978-3-030-22244-4

Table 1. Sample problems from the human physiology ITS.

Scenario:

Mr. C. is walking to the bus stop when he sees the bus. He starts running to catch it, but feels short of breath and has to stop and rest.

Question: *(multiple choice)*

What is the first thing that happened in Mr. C's physiology that caused him to feel short of breath?

Before testing the system with students in the course, we wanted to identify which hints were the most useful so that we can concentrate on the development of those hints. For this purpose we have built a simulator which simulates students randomly receiving the three kinds of hints. Since each hint takes a different amount of time to deliver and has a different probability of success, the simulator is needed to identify the most useful hints.

2 Background and Related Work

Reinforcement learning is a machine-learning paradigm that learns to map situations to actions in order to maximize a numerical reward. The reinforcement learning agent learns which actions yield the largest reward by trying them [1]. In our simulator, the reward for a given action will be inversely proportional to the total time taken by a hint, including both hint delivery time and the time taken by the student to respond to the hint.

This approach is an extension of the approach taken by ADVISOR [2] and AgentX [3]. Both of these systems provide textual hints and learn to provide the one which optimizes student learning. However, neither system provides multimedia hints that take varying amounts of time to deliver.

3 Methodology

To obtain accurate input for the simulator, we needed to estimate how long it would take a student to answer a question with each type of hint. For this purpose we timed three students answering similar short questions. For textual as well as image-based hints, it took the panel an average of 20 s total to both process the hint and provide a new answer. We believe that these numbers were similar because the textual hint referred to the concept map, which students were allowed to refer to at any time during the experiment. In addition to the expected 20 s that it took students to rework their answers, the video clip content required approximately 75 s to deliver.

The simulator uses a temporal difference algorithm to train two agents. We experimented with both the Q-learning formula and the expected value SARSA approach. Q-learning is guaranteed to have early convergence during training [1], but the expected value SARSA approach produced better results overall.

We trained the agents for 1000 episodes using learning rate = 0.25, ε = 0.2 and discount = 1. We chose an undiscounted reward because we were more interested in maximizing long-term rewards. We used an exponential moving average instead of a simple weighted mean to give more weight to the recent total rewards.

4 Results and Future Work

Figure 1 shows the performance of the SARSA and Q-learning methods over 1000 episodes using ε-greedy action selection with ε = 0.2. Although Q-learning takes a lead over SARSA in the beginning, its performance is worse than that of SARSA over time. Both algorithms converged after 200 episodes and provided satisfactory performance.

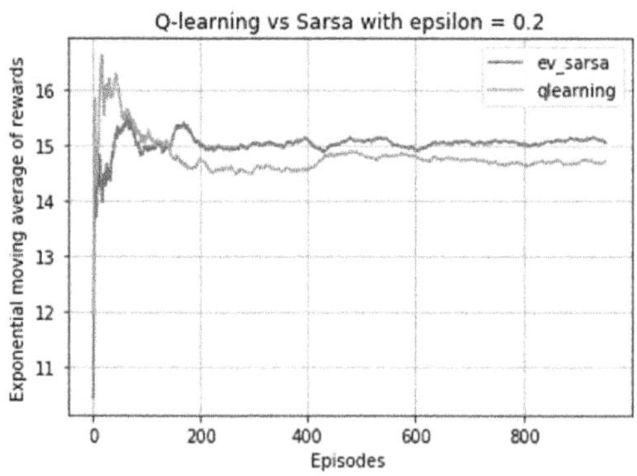

Fig. 1. Results of the two simulation algorithms.

Once the full set of hints has been implemented, we will use the simulator to determine a final policy. In addition, we are looking forward to additional experiments using more complex reward formulas to better model the tutoring situation, and of course to testing the system with students taking the human physiology course.

References

1. Sutton, R.S., Barto, A.G.: Reinforcement Learning: An Introduction, 2nd edn. MIT Press, Cambridge (2018)
2. Beck, J.E., Woolf, B.P., Beal, C.R.: ADVISOR: A machine learning architecture for intelligent tutor construction. In: Proceedings of the 17th National Conference on Artificial Intelligence (2000)
3. Martin, K.N., Arroyo, I.: AgentX: Using reinforcement learning to improve the effectiveness of intelligent tutoring systems. In: Seventh International Conference on Intelligent Tutoring Systems (2004)

Author Index

GPSR Compliance
The European Union's (EU) General Product Safety Regulation (GPSR) is a set of rules that requires consumer products to be safe and our obligations to ensure this.

If you have any concerns about our products, you can contact us on

ProductSafety@springernature.com

In case Publisher is established outside the EU, the EU authorized representative is:

Springer Nature Customer Service Center GmbH
Europaplatz 3
69115 Heidelberg, Germany

www.ingramcontent.com/pod-product-compliance
Ingram Content Group UK Ltd.
Pitfield, Milton Keynes, MK11 3LW, UK
UKHW021019050925
462611UK00012B/1327
* 9 7 8 3 0 3 0 2 2 2 4 3 7 *